Stay Empowered! ♡

Serena

Still Empowered!

Sarah

Empowered

Frame Your Narrative. Own Your Power.

Serena Sacks-Mandel

Aurora Corialis Publishing

Pittsburgh, PA

Empowered: Frame Your Narrative. Own Your Power

COPYRIGHT © 2024 by Serena Sacks-Mandel

All rights reserved. No part of this book may be used, reproduced, stored in a retrieval system, or transmitted by any means—electronic, mechanical, photocopy, microfilm, recording, or otherwise—without written permission from the publisher, except in the case of brief quotations embodied in critical articles or reviews.

For more information, please contact cori@coriwamsley.com.

The author does not represent Microsoft or any of the companies mentioned in this book, nor do her ideas or opinions represent the ideas or opinions of those companies.

Although the publisher and the author have made every effort to ensure that the information in this book was correct at press time and while this publication is designed to provide accurate information in regard to the subject matter covered, the publisher and the author assume no responsibility for errors, inaccuracies, omissions, or any other inconsistencies herein and hereby disclaim any liability to any party for any loss, damage, or disruption caused by errors or omissions, whether such errors or omissions result from negligence, accident, or any other cause.

The advice and strategies found within may not be suitable for every situation. This work is sold with the understanding that neither the author nor the publisher is held responsible for the results accrued from the advice in this book. The content of this book is for informational purposes only. This book is not intended as a substitute for consultation with a licensed practitioner.

Use of this book implies your acceptance of these disclaimers.

Paperback ISBN: 978-1-958481-23-3

Ebook ISBN: 978-1-958481-24-0

Printed in the United States of America

Cover by Karen Captline, BetterBe Creative

Editing by Ben Sacks

Advance Praise

"From the very first moment I met Serena in Atlanta, I was drawn into her absolutely compelling story of navigating life's challenges both professionally and personally. She is a fearless and relentless navigator, tirelessly steering the path for positive outcomes regardless of the circumstances being faced.

"For emerging and existing leaders who want to overcome personal and professional challenges or inspire others to do so, Serena's life lessons in survival, self-sufficiency, and the power to empower others is a must read.

"Kudos to Serena for sharing her very personal look at life and the working world."

~ Mark Pelliccio
Office of the CEO, HMG Strategy, LLC

"As someone who has closely collaborated with Serena and witnessed her unwavering dedication to personal and professional growth, I believe *Empowered* will be an invaluable resource for emerging and existing leaders.

"It holds special significance because it addresses the personal and professional challenges we often encounter on our journeys, especially as female leaders, while also serving as a powerful tool to inspire others. Serena's extensive experience, spanning over 35 years in various industries and roles, gives her a unique perspective that resonates deeply with readers. Through her personal anecdotes and lessons learned, you will gain practical insights on overcoming obstacles, finding resilience in the face of adversity, and embracing the qualities that define impactful leadership.

"Serena's guidance will empower you to navigate the complex maze of your career, instilling in you the perseverance and authenticity needed to thrive. Her book is a gift that encapsulates her journey, filled with triumphs, struggles, and invaluable lessons learned along the way."

~ Julie Young
Vice President of Educational Outreach & Student Services at ASU; Senior Advisor to ASU Prep Academy

"Trauma lingers and lies to us about our worth and capability in this world. This book will be a treasured resource of inspiration not just for those who have experienced trauma, but for anyone who desires to be an impactful leader. Serena is a powerful speaker, and I'm so grateful that her story has now transitioned to book form. The world needs this story. The world needs to see and know the power of forgiveness to heal and create true autonomy and joy in our lives."

~ Andrea Flack-Wetherald
Social Worker, Comedian, Speaker, Author of *The Funny Thing About Forgiveness: What every leader needs to know about improv, culture & the world's least favorite F word*

"*Empowered: Frame Your Narrative. Own Your Power.* is an important resource for individuals striving to identify and/or accomplish career goals and achieve their personal best selves.

"Serena Sacks-Mandel tells her compelling story of overcoming adversity, lack of support, lonely times, and illness to become an outstanding leader in the male-dominated field of information systems and technology. She has overcome challenges thrown at her and has emerged as a strong, empathic, internationally recognized industry leader and speaker. This book will surely inspire others to examine and overcome the roadblocks we all invariably face."

~Selma C. Kunitz
Founder and Former President at KAI Research, Inc.

"Serena's wonderful book is a timely and important reminder that successful leadership requires empathy, understanding, compassion, and strength. In her book, Serena shares her stories of achievement, renewal and career ascent. She also shares her stories of disappointment and heartbreak. That's what makes this a powerful and helpful book. Definitely worth reading!"

~ Mike Barlow
Award-Winning Journalist and Coauthor of *Smart Cities Smart Future*

Table of Contents

Foreword i
Introduction v
Chapter Summaries xi
Empowered 1
    Knocked Down, but not Out! 2
    Not Enough 3
    It's about You, the Reader of my Story 4
    Heather's Story 5
    How I Became the Hero of My Story 7
    Inspiration for the Journey 9
    Key Takeaways 13
Thrive, Don't Just Survive 15
    In the Beginning 15
    Against the Odds 17
    Joseph and the Coat of Many Colors 18
    "What Doesn't Kill You …" 20
    The Second Family was no Better 24
    Functional and Effective 26
    Pandemic 28
    Key Takeaways 32
Education 35
    Lifelong Learning 35
    Leslie 37
    My Future Nearly Stolen 38
    Carpentry and Homelessness 41

    Christine 44
    Tikkun Olam 45
    AI in Education 47
    Key Takeaways 48

Self-Sufficiency 49
    Mid-Century Modern 49
    Early Lessons that Led to Self-Sufficiency 52
    The Dark Side of Self-Sufficiency 55
    Learning to Ask for Help 58
    Caregiving 61
    Interdependence 65
    Key Takeaways 66

Purpose 67
    Motivation 67
    Phase-Aligned Purpose 69
    Contribution 71
    Exhilaration 73
    Serenity 74
    How to Find your Purpose 77
    Adaptation 79
    Relationships Matter 80
    It's not about You 83
    Key Takeaways 83

Perseverance 85
    Credit for my Three Aunts 85
    Family Balance 86
    Career Choice 91
    IBM 94
    Walt Disney World 99
    Harcourt 103
    Hospitality Company 104

    Florida Virtual School 106
    Fulton County Schools 109
    Microsoft 114
    Key Takeaways 116
Resilience 119
    My Biological Father's Death 122
    Layoffs 124
    Divorce 125
    My Stepfather's Death 133
    Cancer 134
    Key Takeaways 150
Leadership 151
    Leadership and Motherhood 151
    Culture of Excellence 155
    Psychological Safety 157
    Structure & Clarity 159
    Shared Purpose 159
    Results 160
    Qualities of Leadership 160
    Assessing Leadership 161
    Leadership Mindsets 162
    Visionary Leadership 163
    Be Inspiring 166
    Value in Feedback 167
    Key Takeaways 169
Triumph 171
    Joy 171
    Success 172
    Soulmates 176
    From Resilience and Strength to Triumph 184
    Serenity 186

Healing Relationships 187
　　Key Takeaways 191
Postscript 193
Key Takeaways from *Empowered*..195

Foreword

Diane Meiller-Cook

Whether you are an aspiring high-powered executive or seeking to test the strength of your soul, Serena Sacks-Mandel's story will inspire you to embrace life's challenges, no matter how formidable, with renewed confidence and fortitude. Like watching a biopic through a high-definition camera lens, the reader sees themselves in Serena's darkest moments, and relives pivotal life moments, witnessing how and why decisions were made.

This book shows us how to persevere with laser focus on the windshield of the future, and reflective glances in the rearview mirror. We learn to drive forward and leave the past behind, with wisdom as the only proof of the road we traveled. Tears drop when she is in pain, and we cheer when she pushes on, both for her success and for our future. When we read her story, we embrace our own as she embraces hers.

Who is Serena Sacks-Mandel? She is a woman who intentionally lives a life full of purpose, awareness, strength, and impact. Like us, her life events were not accidental but the result of experiences that she shares with us throughout her book. She is an achiever who measures the value of her life not by her financial bank account, but by the wealth and distribution of her knowledge that inspires others to embrace life and its challenges with confidence and optimism.

For over 25 years, I've had the experience of being part of Serena Sacks-Mandel's evolutionary journey and have seen the ways she overcame extraordinary challenges. We are friends, business colleagues, fellow executives, and thought leaders. We have worked with and for one another, supported professional and personal dreams, were part of the village in which our children were raised, and challenged one another in a way only trusted confidantes can.

Her challenges started early in life. Many may have seemed insurmountable at the time. As a small child, she witnessed family dysfunction when her parents divorced. This was repeated many years later with her own tension-filled divorce, which temporarily impacted her relationship with her two beautiful daughters who she admires and is so proud of. For years, Serena made progressive career changes requiring 15-hour workdays, while simultaneously suffering a unique life-threatening form of cancer in her early 50s.

Serena never succumbed to being a victim of circumstances but, as she always does, dug deep for inspiration, spread her intellectual tentacles to better understand her purpose, and with reflection, optimistic spirit, and love for life and humanity, she charted her path of success, reward, and fulfillment. Her life's journey took her from extreme skiing on the most challenging slopes, to a remote village in Africa with her daughters to help build a school for young girls, and to Nepal to meet and mentor women with dreams.

Serena seeks experiences that push her physically and intellectually to the highest limits and brings those lessons back

Foreword

into her relationships, whether as a leader, colleague, friend, mother, wife, or daughter. She loves life with the passion of a Mount Everest climber, accepting the drastic changes in its weather, predictability, and surprise encounters. For all who have had the experience of joining her path, whether for a short or a long period, we have been forever touched by her intellectual curiosity, sensitivity, generous spirit, self-reflection, determination, and love. Today, she walks this path with her husband Scott, a partner extraordinaire and mensch, and has reached the pinnacle of life's achievements: self-love, self-worth, and self-confidence.

Introduction

Most people I know have wondered whether they were making the right choices and doing the right things at some point in their lives. Many times in my life, I questioned whether I was on the right path—if all the struggle could and would someday pay off.

It's difficult to see around the corners of our lives. When we make one choice, we never know how life would have turned out had we made a different choice. Choices often seem limiting—making one decision eliminates the possibility of another. I've experienced this limitation at times, like when I decided to pursue a career in technology at IBM instead of accepting the Presidential Management Internship and starting a career in government.

However, sometimes, we can change our choices later. For example, after working for corporations for 25 years, I pivoted to public education, and now, after 10 years as a CIO there, I am back in the corporate world. A decision early in a career puts someone on a certain path, but a later decision to pivot in a different direction can always be considered. A book like this would have helped give me confidence to follow my heart, pursue my dreams, and not worry so much about the outcome early in my journey.

I wrote this book to inspire and encourage others and to give them the hope and direction I needed throughout my career. When I started my career, I knew the journey was going to be long and challenging, but the challenges I could anticipate were not my only challenges. As I traversed my path, I experienced what could have been crippling difficulties along the way without anyone to support, guide, or encourage me. I raised my children while working in an extremely demanding role and while managing a difficult marriage and then divorce. I moved to a new city and established a new professional network three times and was diagnosed with cancer twice, undergoing treatment by myself for two years without taking a break from my professional and nonprofit work. Not to mention, I was often the only woman in a room of technologists, or the only woman at a leadership table far too many times to count, even within the last 15 years!

Nonetheless, I've persevered through my personal and professional challenges and have been grateful to grow while experiencing an incredible breadth of experiences as a leader, consultant, board member, individual contributor, and intern. I've worked at Fortune 100 companies, nonprofits (social value-based organizations), boutique firms, the public sector, and educational organizations. I've also been blessed to find acceptance and serenity throughout my battle against cancer, which ultimately helped me open my heart again, or perhaps for the first time, and allowed the blessing of love to enter my life.

I feel compelled to tell my story with the hope that it will inspire my contemporaries and support the next generation of leaders. Beyond that, I sincerely hope that this book can benefit any woman who is an emerging leader who wants to overcome personal and professional challenges or who wants to inspire others to do so.

As leadership, especially in the tech realm, continues to transform, it's clear that there is still work to be done to empower women in their professional journeys. Imagine the impact if colleges and career centers offered more robust support for career development. Navigating the twists and turns of a career path is no cakewalk, and having additional guidance is key. Women in male-dominated fields grapple with unique challenges, and it's important for everyone, including men, to understand and support each other through these obstacles.

We may be past the era of female and minority tokenism at the team member levels, but we have a long way to go at the leadership level. Most cabinets and boardrooms are still dominated by men, and certain fields like technology have a disproportionate ratio of men to women.

For women of color, the statistics are alarming. According to McKinsey, "The representation of Black, Latina, and Native American women in tech jobs is declining despite their rising share of tech degrees. Inclusive policies and practices could help companies retain them."[1] I believe that women need more inspiration, support, and stories of success to show them that it is possible to rise up in technology and be successful on one's own terms.

This book is especially meant to inspire and encourage emerging female leaders in technology, science, engineering, and

---

[1] "Empowering Black, Latina, and Native American women in tech." McKinsey & Company. Aug. 29, 2023.
https://www.mckinsey.com/industries/social-sector/our-insights/empowering-black-latina-and-native-american-women-in-tech#/

all fields where women are under-represented. Women still need to overcome too many hurdles to be seen and heard. Our voices and ideas are needed to represent the female perspective.

My story is a triumphant journey, and yours can be too. When I shifted my focus to my purpose and intentions, I discovered a newfound strength in my core, understanding who I am, why I'm here, and how I want to live. This inner confidence coexists with a continuous willingness to learn and grow. Challenges and frustrations persist, but my ability to handle them has improved, and I bounced back more swiftly. The need to prove my worthiness has faded, and I now embrace my truth. I am enough.

My purpose is not about achieving wealth, fame, or accolades. It's about contributing to others' lives and making a positive difference in the world. What success means to each of us is as varied as the number of people reading this book. For me it means transforming education with technology and taking care of the people I love. That may sound simple, but it's been a long, challenging, and winding journey to get here. Hopefully my story will encourage you through your own unique journey.

I have presented aspects of this book to live audiences several times, and the result has been phenomenal; many people have approached me or written to me about how my story touched them. I have found that my stories create connections, acceptance, and understanding. Hearing how you feel about my work and my writing is the greatest gift I have received, and I hope to continue to positively impact lives.

A quote that resonates deeply with me is often attributed to Nelson Mandela, but was actually written by Marianne Williamson:

Serena Sacks-Mandel

Our deepest fear is not that we are inadequate.
Our deepest fear is that we are powerful beyond measure.
It is our light, not our darkness
That most frightens us.

We ask ourselves
Who am I to be brilliant, gorgeous, talented, fabulous?
Actually, who are you *not* to be?
You are a child of God.

Your playing small
Does not serve the world.
There's nothing enlightened about shrinking
So that other people won't feel insecure around you.

We are all meant to shine,
As children do.
We were born to make manifest
The glory of God that is within us.

It's not just in some of us;
It's in everyone.

And as we let our own light shine,
We unconsciously give other people permission to do the same.
As we're liberated from our own fear,
Our presence automatically liberates others.

Empowered

~ Marianne Williamson[2]

The group I trekked with in Nepal with Mt. Everest in the background. We are by the Edmund Hillary statue—he was the first to summit Mt. Everest.

---

[2] Williamson, Marianne. *A Return to Love: Reflections on the Principles of "A Course in Miracles."* HarperOne. First edition, 1992.

Chapter Summaries

I broke this book into chapters based on major themes in my life.

Empowered: Empowering others has been one of the most rewarding aspects of my life. I'm never happier or feeling more exhilarated than when I've helped someone else to be successful. Being empowered is about owning your power and helping others to own theirs.

Thrive not just Survive: At times in my life, I felt helpless, attacked, or bullied. Sometimes it was accompanied by loneliness, which led me to feel hopeless. These were pivotal times when I chose whether to simply survive and let my circumstances shape me, or to thrive and shape my circumstances.

Education: Lifelong learning is essential to staying relevant and active in today's rapidly evolving world. It is important to always have a growth mindset. I've found that confidence comes with knowledge. Lifelong learning allows you to have a success mindset… don't ever let anyone steal that mindset from you! Learning takes endurance, resilience, and determination. Find mentors that help push your limits.

Self-Sufficiency: This is the mantra from my early years, and it was drilled into me that I had to be self-sufficient to make my own way in this life and not depend on anyone. This turned out to be a double-edged sword.

Purpose: It has always been important for me to have direction, to stay motivated, and see meaning in my activities. The way I see it, my life was given to me as a miraculous opportunity to achieve meaning. I've felt a calling to do work that improves lives and makes a difference for others.

Perseverance: Perseverance is the quality of continuing to work toward a goal despite obstacles or setbacks. Sometimes perseverance can be your most valuable tool. I try to maintain a long-term focus, set achievable targets, and stay motivated throughout my journey. It is not enough to know your purpose; it must be pursued relentlessly.

Resilience: Resilience is our ability to withstand adversity and to bounce back when we get knocked down by it. It has been my number one strength and has helped me overcome challenges. I have been given plenty of opportunities to develop and demonstrate this skill. Besides horrific formative years, I was struck with rare and aggressive uterine cancer in my early 50s. I endured multiple surgeries, a year of aggressive chemo, and other treatments through which I never stopped working, only to plunge into depression for another year.

Leadership: Leadership is not a position, title, or role. It is a set of qualities and disciplines that are practiced every day, including influence, integrity, value, vision, curiosity, and inspiration. As a leader with a team, it is important to be authentic, set clear expectations for your team, know, and leverage each person's strengths, and help them develop their skills and experience.

Triumph: Triumph is the outcome of framing your narrative and owning your power. It is the culmination of empowerment, survival, persistence, education, resilience, leadership, self-sufficiency, and purpose. This is when all of one's efforts come together and yield success on one's own terms. For me, I feel triumphant when I am helping to transform education with technology and when I'm caring for the people in my life who I love most.

Empowered

Being empowered is about framing your narrative. People will try to tear you down just because they can. They will try to take your power away from you just because they can. You can't stop what people do, but you can write the story you tell yourself, and you can determine how you respond to it.

How many times have you been the only woman in the room at a technology meeting or the leadership table? If you are a woman in technology, you are one among far too few, and if you are a woman of color in technology, you are among even fewer. I believe that women are not staying in technology and achieving leadership positions because they do not have enough support and inspirational role models.

I may appear to have many advantages — white, middle-class family, educated parents, grew up in Westchester, New York — but you will see in this book that I had tremendous challenges, which could have knocked me out of the leadership and technology race. Just as I overcame the odds, so can you. You can become the hero of your story, not the victim. In doing this, you claim your power and can use it to empower others.

You too can be a role model for others by sharing your story and experiences. You can inspire and empower others to pursue their dreams and achieve their goals. Remember, you are not alone, and there are many people who are rooting for you!

Knocked Down, but not Out!

I'm going to tell you a story that I've used as a personal driver throughout my career. While I've had much support and guidance, I've also had experiences where people tried to tear me down. At the end of my senior year of high school, I felt knocked down and unsupported. I had been accepted to attend my top choice school, a prestigious private university. My father was well-off financially and had paid the tuition for my three older siblings. Although we had a tumultuous relationship, he had confirmed with me during high school that my college funding was secure regardless of where I decided to go.

But when the time came for the tuition deposit, instead of a check, I received a phone call from my father who told me that he would *not* be paying for college after all. Not because he didn't have the money, but because he didn't want to pay. He simply did not like me and accused me of being a "bad" daughter. He didn't care about my excellent grades, varsity sports, or work ethic at my job. It was about my loyalty to him versus my mother after their messy divorce.

This was a blunt-force attack that could have affected me in devastating ways, but I chose to use this experience as fuel. My stepfather helped me find a way to work out the finances and attend a state school, but I was determined to show my father and the world that I was worthy of college. I was capable and ready for the challenge. Every "A" I received in college, every bonus and promotion were evidence that I controlled my destiny and that no one could steal my power.

Years later, in my performance review, my first manager at IBM said, "Serena is incredibly resilient and needs mountains to climb." This female leader of mine nailed it, and I've been bouncing back from setbacks ever since … while finding higher mountains to climb.

How did I turn this emotional blow into victory? I could have said, "My father stole my future. He wouldn't pay for college because he didn't believe in me. I must be a failure." Instead, I framed it with self-empowerment. My story is, "My father tried to steal my future, but I succeeded despite him." The reality, that I went to a New York State school instead of the private university, is the same either way. The mindset makes the difference.

Not Enough

I've had wonderful mentors who have built me up, and I've also encountered those who tried to tear me down and tell me I wasn't good enough. My success in college was not enough. I still felt that I needed to show the world, especially my father, that they had been wrong about me. I felt capable, yet I still wanted affirmation. I promised myself that I would be more than merely self-sufficient; I would be successful on my own terms.

My father married again, for a third time, and I went to visit during a break from my freshman year at college. I wanted him to see that I was doing well. I wanted his affirmation that I was taking on the challenge of college and succeeding. I wanted him to finally *see* me and my potential. But wanting and having are two different things; the visit turned out to be a disappointment.

I had to come to terms with the fact that my father would never truly see or accept me. He and his new wife ignored me that day, fawning over their newly adopted baby, who was much more important to them. He never gave me a chance to describe my successes, and once again, I felt like I didn't matter to him at all. I left that house knowing it was more important for *me* to see clearly who I was and trust in that. That was a step in the right direction.

I wanted so badly for him to give back the power he had taken from me, only to learn that *I* was the one who had to claim my own power. He had absolutely nothing to do with it. The song that I used in an eighth-grade slide show project kept coming back to me over the years: Whitney Houston singing "The Greatest Love of All." She was so right. I had to love myself; that was most important. He didn't see that "children are the future" or know how to "teach them well, let them lead the way," but I did not need him. I only needed my self-love and hard work. I was even more determined to get my education and then to succeed in my career. This may have been the first time that I realized that I had to be the hero of my story, there was no one to take care of me, no one to save me, no backstop when the going got tough, as it always does.

It's about You, the Reader of my Story

To me, being empowered is not just about finding and owning your own power; it's also about helping others to own their story. Empowering others has been one of the most rewarding aspects of my life. I'm never happier and I never feel more exhilarated than when I've helped someone else to be successful.

Empowering others is a process. First, we must empower ourselves by accepting what's happened and by framing our narrative as the hero, not the victim. Next, we must forgive those who have hurt us: to heal ourselves, not because they were right.

I've had lots of forgiving to do in my life—I share more on those painful and traumatic moments of my life in the next chapter.

We must take ownership of our power, take responsibility for what has and has not happened. *When we confidently own our power, we naturally support and empower others.* For me, enabling the success of others creates a feeling of elation, great fulfillment, and exhilaration. This has been a common theme throughout my life.

"True power is the power to empower others."
~ Che "Rhymefest" Smith[3]

This quote beautifully encapsulates the idea that genuine strength lies not in dominating or controlling others, but in uplifting and enabling them. When we empower those around us, we create a positive ripple effect that extends far beyond ourselves.

Heather's Story

While I was working as a consultant, I met a young, intelligent, hard-working woman who in some ways reminded me of an earlier version of myself. At the time, she was raising two young girls as a single mother. She was struggling with all the issues single moms must deal with while managing a low-paying position. She came to work late one morning because the

---

[3] https://wmfpodcast.org/rhymefest/

baby had thrown up, and she had been up all night with the older child and was managing a migraine headache too.

I know from experience that raising two very young children wasn't easy *with* a partner, but alone, it was all on her: meals, baths, illnesses, taxiing kids to school and daycare, not to mention paying for it all. I was immediately impressed by her drive, determination, and ability to get stuff done. I noted that even without any formal legal training, she was far better at handling and negotiating contracts than others in her department. She was overlooked because of her age and underestimated because of her physical attractiveness. Her supervisor didn't appreciate her value.

Through several discussions, we determined that a career as an attorney would be rewarding for her. I arranged a meeting with an attorney and law professor to give her more details. She was sold on the idea, and quickly moved to get her qualifications and applications to attend law school.

Even if you're not caring for two young daughters and continuing to work a full-time job, law school is a grind. She had the energy of youth on her side, though. Still, I'll never fully understand how she managed all those responsibilities and time constraints and still graduated at the top of her class. Many young lawyers struggle even after graduation. But she quickly found her stride with the same dedication, resilience, and caring for her clients that she had put into her studies. We've stayed in contact over the years, and I've followed her progress. I'm continually amazed by how she's been able to be a successful lawyer and mother who continually gives her clients an extraordinary level of service.

Ten years after Heather graduated from law school, I was able to witness her talent and passion in action when I was being unfairly treated and threatened. I was glad to have her on my side, defending me. I never dreamed that a person I had mentored and encouraged would be there to guide, defend, and encourage me later.

Empowering others is always rewarding, and I am grateful for that opportunity. I watched this young person become an accomplished attorney with two amazing daughters who are now pursuing their dreams with her support. She is a true example of self-empowerment and an inspiration for her gifted daughters.

I would like to think that I helped her become self-empowered and realize her dream of becoming a lawyer, where she helps others improve their lives and achieve their goals. Each one of us has the ability to help others. Our time, perspective, and friendship all have value. When we feel appreciated by someone else, our sense of empowerment is real; it gives our lives purpose and meaning.

Heather is a testament to framing her narrative and taking responsibility instead of blaming others. She owns her power, supporting her daughters, traveling the world, and being an amazing role model for them.

When we own our power, we empower others. That is true power and fulfillment.

How I Became the Hero of My Story

It has not always been easy for me to look at my history and frame myself as the hero. The hero authors their story and takes responsibility for the success or failure of the outcome. It has taken many pain-filled years, an enormous amount of reading, self-reflection, therapy, and workshops to overcome my self-loathing. A healthy mindset did not come naturally to me; I've done the work to get here.

There were countless nights that I cried myself to sleep. There were lonely birthdays, many bouts of emptiness, doubt, and hopelessness. Depression and anxiety were also in the mix. However, no matter how bad things seemed at the time, I found that it was always best to struggle through and hang on until things improved, because they almost always do. The message from the Torah reading at my bat mitzvah was, "Things happen for a reason, and everything works out for the best, even if we don't see it at the time." I needed to take that to heart.

To frame our narrative positively, we must gain perspective. It's imperative to step away from anger, hurt, and pain by processing these emotions instead of burying them. Those feelings and emotions are real and meant to bring our attention to something. The next time you feel anger or pain, ask yourself, "What do I need to see, hear, or learn from this difficult situation?" Instead of asking, "Why is this happening to me?" you can reframe the question as "What can I learn from this situation? How can I grow from this?"

I completely understand that this is tough, but give it some time for the emotions to mellow. Then go back to the questions that will propel you forward, not hold you back. What I try to do is process emotions by naming them, feeling them, listening to what I need to learn, and then letting everything go.

In the end, we are not our history; we are what and who we decide to be. Some people decide more overtly, and others decide by not making a choice, just letting life happen to them. But either way, it's a choice. I choose how I show up, what I do, and who I am. To be who I am, I can't live in the past. I need to learn from the past and forgive those who attacked me to continue moving forward toward my goals.

Inspiration for the Journey

The stories in this section speak about my journey. Some of the stories include people who have inspired me, and I hope they inspire you too.

Briana

My oldest daughter started higher education at a large university and found it didn't mesh well with how she learned. She saw that the typical classroom situation wasn't working for her and decided to follow a nontraditional approach. She called an online meeting with her father and me and shared her PowerPoint with us, detailing her budget and long-term plan. (I like to say, "She had me at PowerPoint.")

Her plan was to go to the largest, busiest city in America, enter the fashion industry, stay in school, and finish her formal education online while getting a real-world education. This is a plan that most people wouldn't be brave enough to start let alone finish successfully, but sometimes a parent must put aside their fears and let their child find their way.

My daughter has always had grit and determination, and I knew if she believed in this dream, she would make it come true. My job here wasn't to advise; it was to encourage and empower. There were plenty of people who didn't believe she could be successful, and I would absolutely not be one of them. She had a chip on her shoulder from the naysayers and used that energy to climb her mountain. She learned about the fashion industry, knew the designers, understood the styles, and found her niche in outfitting the elite of New York City. She continued to work in retail and advanced in her career until March of 2020, when the entire world shut down for the pandemic.

Instead of sulking, she went to the store, purchased a whiteboard and remapped her future into tech sales. Through grit and determination, she once again quickly advanced and was offered a prestigious executive sales program that was only offered to three people out of thousands of candidates. Her next goal is to get a master's degree in business from one of the most prestigious business schools in America. I have no doubt that her future holds mortarboards and hoods in the color of her school. I know that I will be there to encourage and empower this courageous young woman. She is an example of forging your own path, finding your power, and owning it.

Briana's grit and determination remind me of my mother. My father did not like the bids he received for paving the driveway, which was very steep and a quarter mile long. My mother knew it could not remain in its dirt and rock form. So, she rented a steamroller, contracted for the blacktop material and paved the driveway while my father documented it on 8mm film. We loved watching this footage at family Thanksgiving gatherings. I'd like to think she passed this can-do attitude down to her daughter and granddaughters. Owning our power is often about breaking down barriers and stereotypes, and not accepting the status quo.

Mickey

My mother-in-law is a reminder of what an empowered woman can do. She broke down barriers in the 1950s doing stand-up comedy. After having children, she used her love of baseball to write a local column for the Chicago Cubs. At the time, no one would have read a column from a female sportswriter, so she used her androgynous nickname, Mickey, to byline her column.

In 1969, she realized that the family needed additional income, so she started and ran the family business that would be the family's sole source of income for the next 30 years. The family business started out as a taxi company, and my mother-in-law was the first driver, dispatcher, and bookkeeper. Her husband was fired from his job as an insurance agent because his company refused to believe a woman could run a cab company.

While running her family business, she never gave up her love of entertaining and directing. She continued to produce and star in shows at a local theater. Today, at 86, she travels internationally on her own, multiple times per year. Last year for her 85th birthday, she stepped foot on the continent of Antarctica. I love that she owns her power daily, and that my husband often can't get a hold of her because she is out running some event.

You Won't See Me on *America's Got Talent*!

Being my authentic self means listening to my inner voice and enjoying the things I love, regardless of what others think of me. I love music. My first concert was in high school, and we saw The Police. I was hooked. Live music, dancing, singing along with the band — that was my jam! But through the years, people pointed out that graceful dance moves didn't run in my family, and crooning was not my gift. My singing voice slipped away because I listened to those who reminded me that I was no Whitney Houston. My dancing stopped as I was told I looked more like Elaine on Seinfeld than Madonna. It was insidious; it didn't happen in a moment or a single comment. It chipped away over time, and slowly disappeared.

I started to convince myself I couldn't dance, I couldn't sing and, even worse, that I shouldn't do those things. This great joy in my life had slowly slipped away until one day, driving to Tennessee, Scott (my fiancé at the time, and now my husband) was singing in the car and noticed I was just barely mouthing the words. He prodded, pushed, and poked at me to sing along, but years of training made me resist. He never gave up and continued encouraging me to sing with him!

Because I had agreed to marry him, I gave in to humoring this strange request. Slowly and surely, I found my voice. And though our collective sounds, I'm sure, make dogs miles away howl in agony, I found my joy in singing again.

Somehow this just wasn't enough for this crazy man, and he insisted we dance. We now not only spontaneously dance around our family room; we have dedicated '80s dance nights, where there are no holds barred on whatever moves we dream up. These moments of song and dance exhilarate and empower me. They are a reminder that I own my power and my story when I am my true authentic self, even when that means singing off-key and being silly.

Laughter is healing, fun, and rejuvenating. I am grateful to have a friend and partner who reminds me to find my joy, listen to my inner self, and own my power, and he makes me laugh till it hurts sometimes.

Key Takeaways

- Frame yourself as the hero of your story, not the victim.
- True power is the power to empower others.
- Process your emotions. Name them, feel them, listen to what you need to learn, and then let them go.
- You are not your past; we are what we decide to be.
- We get to define success in our own terms.

Thrive, Don't Just Survive

I'm the youngest on the left, with my three siblings.

My youth consisted primarily of learning and practicing survival tactics. I could have been a statistic and turned to drugs, gave into depression, or worse. Somehow, I managed to overcome the traumas I faced and thrive through education, hard work, and trusted mentors.

In the Beginning

Even though I was the youngest child of four, my siblings called me "Grandma" because I held everyone accountable for our parents' strict rules … or maybe they were just teasing me.

Empowered

Perhaps I was difficult, rambunctious, and strong-willed. These traits that were not celebrated in children's behaviors at the time are now seen as early leadership traits! For better or worse, my attitude was, "You didn't want me, but I am here now, so deal with it. I'll show you!" This chip on the shoulder can be the right fuel to weather storms, and that I did. I survived my younger years but was far from thriving. It was a war zone inside the family, and I had to be tough to become who I am.

There have been times in my life when I felt alone, helpless and bullied. Though difficult, these situations were pivotal to my development. It is in the midst of personal and professional challenges, when our strength and resolve are tested, that we become our best selves.

I may not have had much control over my original family circumstances, but I chose whether my circumstances shaped me, or I survived and thrived in shaping my life. We can't always change what happens, but we can control how we respond. Each time I hit a roadblock in my path, I went around or over it and survived. It wasn't always evident in the moment that the outcome would be alright, but in hindsight, with each survival moment, I became more prepared for the next. Success builds on success. I began to thrive when I stopped doubting myself and hesitating when making the decisions I needed to make to survive.

This is the essence of resilience: the ability to bounce back from adversity, to adapt and thrive in the face of challenges, to turn setbacks into opportunities. Resilience is not a fixed trait that some people have, and others don't. It is a skill that can be learned, practiced, and cultivated. Resilience is not about avoiding or denying difficulties, but about facing them with courage, confidence, and creativity. My resilience has been tested since before I was born.

Against the Odds

I am not supposed to be here. That's what my parents told me, and their words echoed through my childhood like a haunting refrain. They never intended for me to exist—my conception was an accident, a glitch in their plans. The contraception failed, and I emerged, an unwelcome surprise.

Why did they share this painful truth with me? Perhaps it was anger, frustration, or a desperate attempt to justify their own struggles. Whatever the reason, their words etched themselves into my young mind. At six years old, I carried the weight of their failed intentions, a burden no child should bear.

Our home was a battleground—a war zone of divorce proceedings that stretched over thirteen grueling years. Legal settlements dragged on, and I became their pawn, caught in the crossfire. My parents, both brilliant minds, successfully navigated Ivy League institutions, yet their emotional intelligence remained stunted. They were ill-equipped as parents, and I paid the price.

I was no longer a child; I had to be an adult to survive. My siblings escaped much of their wrath, but I was the target. Perhaps they blamed me for their own shortcomings. Regardless, I learned to argue at their level, to fend for myself in a world where safety and love were scarce commodities.

In hindsight, their words fueled my defiance. "I'll show you," I thought to myself. Their dysfunction became my catalyst for growth. I refused to let their truth define me. Instead, I channeled it into resilience, determination, and the will to find my life's purpose. I would forge my own path, create my own narrative — one of resilience and triumph.

The child they never wanted became the person they had to rely on in the end. Success, I realized, builds upon itself. Each survival moment, each small victory earned, propelled me forward. I was not a victim; I was the hero of my story.

My name, "Serena," is the root of "Serendipity," which is the name of a wonderful ice cream shop in New York City and the title of a beloved romantic film about fate and destiny. When I saw that movie as an adult, it occurred to me that my birth was actually a fortunate surprise, not a mistake. Merriam Webster's definition of serendipity is, "The faculty or phenomenon of finding valuable or agreeable things not sought for."[4]

Joseph and the Coat of Many Colors

Domestic abuse and family trauma cut across education, wealth, and race. Our family proved this point. My early years were marked by neglect and the absence of safety, stability, or security. I didn't feel like I belonged anywhere and barely survived the compounded emotional traumas. Therapy and counseling have helped to provide insight, guidance, and a different perspective, but that came later. First, as a young person, I just had to survive.

---

[4] Merriam-Webster Dictionary, s.v. "serendipity (n.),"
https://www.merriam-webster.com/dictionary/serendipity

My siblings treated me cruelly. They were mirroring our parents' stressful and abusive relationship. They did not know better at the time and have apologized since and throughout our adult lives. They regretted what they said and did to me as a child. It took me many years to forgive them. As a child, all I could do was survive and learn to move beyond my circumstances. I share this with you because I meet people all the time with the same story. I want to pass on to you the tools I used and still use to survive and ultimately thrive during these difficult times.

In the end, the story of my siblings turned out much like the Bible story about Joseph and the Coat of Many Colors. You may have seen the same story in the Broadway show *Joseph and the Amazing Technicolor Dreamcoat*. That was my favorite story as a young child because I knew it was going to be my story. My siblings were cruel to me; they didn't want me to be a part of their family, not because I was favored or they were jealous, but because my parents did not want me either. I became the focus of their discontent.

Their rejection of me was akin to how Joseph's brothers left him to die in a pit and told their father he was dead. Joseph didn't accept the story his siblings were writing for him, and I refused to accept the story my siblings tried to write for me. Joseph made a success of his life and, in the end, saved his family. That was not my plan, but it did become my reality. Joseph saved his brothers because of who he was, regardless of how he was treated before. Despite their earlier rejection and abandonment, he did not hold a grudge against his brothers. I used this example and did the same thing.

Empowered

I love my siblings and feel that it is a privilege to hold together what's left of our family and help make each of their lives better. That's why, when they had times of financial and emotional struggles, I was there to support them. I wanted to give them what they needed to reduce stress and create security in their lives. Now, we are connected by the tragedies we endured and have overcome.

Often, people who are hurting say and do cruel things to others. We don't always have the choice to avoid their pain, and the results can be so damaging that it leaves emotional, mental, and sometimes physical scars. Having suffered abuse, I have deep empathy for others who have faced it.

"What Doesn't Kill You ..."

My earliest memories are etched in trauma—the kind that seeps into your psyche and shapes your very existence. Before the age of five, I witnessed my father try to kill my mother three times. I have found that witnessing brutality can leave deeper scars than being the victim of cruelty.

The first time, I was an infant, too young to comprehend the world, yet old enough to witness the brutality that unfolded within the walls of our home. The "shoe heel incident" remains vivid—a snapshot frozen in time. My parents' bedroom became a battleground, and I, an unwitting spectator. My father's rage knew no bounds; he wielded the heel of a shoe against my mother, leaving her battered and broken. The emergency room became her refuge, and my cranial neurons unwittingly stored this horror.

But it was the stairs and the brass gate that haunted me most. My father propelled my mother downward, her life hanging by a thread. Miraculously, she clung to the rail, defying gravity. The third time, it was her neck—his fingers closing in, choking the life out of her. She survived physically, but the emotional scars ran deep. She would cringe when anyone touched her neck, even gently.

Divorce loomed, but societal norms of the '60s kept them tethered—a facade of family unity masking the chaos within. After fourteen years of marriage, abuse and infidelity took its toll, and they divorced. Even after the legal paperwork was complete, their struggle dragged on—a messy, elongated battle that ensnared me and my siblings as unwitting pawns. My parents' animosity knew no bounds; they pitted us against each other. I navigated the minefield, my innocence shattered by their relentless feud.

My mother was abusive and neglectful too. Understandably, she was depressed and angry after the divorce. She was vulnerable too. Her choice in men did not improve for many years. She allowed boyfriends to claim her for the summer and ship us off to sleepaway camp for months at a time. What should have been wonderful experiences were tainted by deep feelings of abandonment.

When we were home, she yelled at us, hit us, and threatened to make it worse. My sister remembers being too afraid of her outbursts to invite friends over to play. My earliest memory was of my mother suffocating me with a pillow when I was a toddler. She nearly killed me and later told me, "Well, I didn't suffocate you, and you are still here!" The implication was that it wasn't that bad and now everything is fine, although it wasn't.

Empowered

Even after the legal divorce, my parents continued fighting in court for the next 13 years. Evidently, my father refused to carry out his commitments in the settlement agreement. My mother, who was without a lawyer due to her financial situation, tried to hold him accountable. Meanwhile, my father, with his high-paid lawyers, wanted to end or reduce his support for us. He fought for custody and lost—we were told he wanted custody to reduce his financial burden, not to be with us more often. At one point, we were told we would have to testify that we wanted to live with my mother. I was scared to testify and feared if I said the wrong thing I would have to live with my father. The court-ordered compromise was forced visitation with our father every other weekend.

They each would tell me how awful the other was being. I internalized these messages and wondered, *if I were the progeny of two awful people, what did that make me?* Even as a child, I needed to survive these messages and tell myself I don't have to be them. This is an ongoing demon that haunts me even today, and I still need to remind myself I survived the past and that it is not my future.

The court-enforced visitation with my father was a recurring traumatic event in my childhood. Neither one of us wanted it, but the court had mandated it. For the most part, I would have preferred to stay with my mother, but young children don't have a voice before the judge.

These visits were bizarre. My father was hardly present. I was left with people who didn't want me around, while my father went to the club to play golf. I spent the time with my stepmother and stepsiblings, sometimes at their large home, and other times at their posh country club, where my father was the club champion golfer. He drove a Mercedes; his new wife drove a BMW. There were so many stated rules and unspoken protocols. It's one thing to grow up poor surrounded by people in the same situation without really knowing you are a have-not. It's another thing to be poor surrounded by wealthy people all looking down on you.

I was embarrassed and ashamed to only own one pair of hand-me-down jeans and one pair of shoes that only got replaced when I outgrew them each year. When my father's family went to the club during a weekend visitation, I was forced to go with them and to mingle with the rich and richer, with that "less-than-them" feeling confirmed by overt glances and scornful remarks. I was not allowed to take tennis lessons because my clothes were not pure white with designer labels (thank you, Serena Williams, for finally breaking down those barriers). I did not have what they had. I was ashamed that my dime store underwear did not say "Bloomies" on the rear and didn't want to change into a bathing suit in the women's locker room in fear that someone would see them. It made me feel sad and worthless. I decided I would earn my own way and buy what I wanted as soon as I could so I wouldn't be dependent on anyone.

And so, I learned resilience — the kind forged in the crucible of dysfunction. Survival meant mastering the art of diplomacy, of walking the tightrope between warring parents. It was not until I was 18 and there was no more leverage that the battle finally subsided. But the fighting left us all tattered and injured. I survived the trauma, and it made me stronger.

The Second Family was no Better

My mother remarried when I was in high school. High school was already hard for me, and my step-siblings made it worse—as if that stage of growing up isn't already difficult enough! Over the summer between my freshman and sophomore years of high school, my mother and I moved into her new husband's home with four of his children. My new stepsiblings were jealous and mean to me, especially his daughter, who was in the same grade. We'll call her Donna.

Donna was loud and argumentative. She flaunted her boyfriend—he'd asked me out first, and she discouraged me from dating him. She bossed her younger brothers around mercilessly. Over the next few years, Donna stole from me, lied to our parents about me, and tried to manipulate me. Donna hit me and hurt me physically too. My stepfather saw that there were issues but refused to hire a counselor because it would be on his business medical records and could hurt him at work.

During this time, all I could do was hang in there and survive. I tried reconciliation with Donna, but it was never successful. I had gone from one bad situation to another. Even as adults with children, I tried to mend fences with her, but then realized this was never going to happen. It can be easy to hold a grudge, but it will bring anger and bitterness into new relationships and experiences. Forgiveness brings a kind of peace that allows you to focus on yourself and helps you go on with life. Eventually, I had to forgive her and move past the trauma the two of us endured together. I'm sure I had my part in our bitter battles, and hopefully she can forgive me too. This was yet another lesson for me that, to survive, I had to let go of people that held me back or tore me down. If you want a good book about forgiveness, my favorite one is *The Funny Thing About Forgiveness: What every leader needs to know about improv, culture, and the world's least favorite f word*, written by my good friend Andrea Flack-Wetherald.

I wasn't exactly my own best friend either. Looking back, I'm not sure how the girl I was in high school survived to make it to college and find success. I was going down a bad path and indulging in unhealthy activities. If I had continued on that path, I would have likely become a statistic: a high school drop-out, in jail, dead, or a struggling teenage mom. I turned my story around, and though I came close to one or more of those things being my future, I managed to make sure it was not. Caring teachers, guidance counselors, my aunts, and my mother all helped me through it in the end. By the time I was in high school, my mother had changed her circumstances too. She and I had grown close, and she helped me pull it together and survive those years.

I knew that it was important to keep up my grades and work hard. Little by little, I reduced the bad behaviors on a regular basis, and somewhere in there, I turned it around. The rebellious 16-year-old gave way to practicality; I knew I would be on my own soon, and I had to prepare for college and my future.

Despite my focus and drive, the residual feelings of not belonging and PTSD from these traumatic years followed me for a long time and affected every relationship I had going forward. It took me years of therapy to put what happened in perspective. I found a therapist who empathized with my story, and for the first time, I felt validated and seen. She said what happened to me was horrific. How they treated me was not right or fair, but it was not my fault either. This revelation changed how I saw myself and my parents. I recommend everyone find a good therapist in their 20s, if possible, and sort out their childhood to enable them to fully engage in life as an adult. It's worth it.

Functional and Effective

My primary survival skill was to not feel anything. I blocked the world physically and emotionally. I simply charged forward through school and my day-to-day life trying to be functional and effective, which became my mantra during challenging times. It saved me, and it served me in those early days. But years later I saw how being so detached damaged some of the relationships that meant most to me. Sometimes, certain things serve as useful survival tools when we're younger, but they need to be re-examined as we mature, as they may be holding us back. In business as in life, Marshall Goldsmith's book rings true, "What got you here won't get you there."

Compared to my siblings, I wasn't the cutest or the smartest and not even the most talented, but I was persistent, resilient, and confident. I used challenges as fuel to move beyond my circumstances. Perhaps I was born with *chutzpah*, the Yiddish word that means "spunk, scrappiness, and nerve." Sometimes I may have more confidence than I deserve, but it serves me well. No one accuses a man of having too much confidence! The good news about confidence is that it can be learned and earned. Small successes lead to bigger ones, and confidence emerges.

Seeing my setbacks as part of the learning and growing process added to my confidence rather than subtracting from it. It wasn't easy; in fact, I looked for the tasks that no one else wanted to take on—the hard ones—so I also earned the confidence of my colleagues. The most important lesson here is to never quit, continue challenging yourself, and believe that, through hard work, you can achieve your goals. I may not be the smartest, the fastest, or the most coordinated, but I can choose to be the hardest-working person in the room.

The good news is that confidence can be learned and earned. For me, learning confidence started when I was very young. It was through these small successes that I built confidence. Later, confidence came from excelling in school and sports. These small successes led to larger successes, and I continued to challenge myself. I was learning from each setback and mistake and gaining confidence from each success.

I adopted the attitude:

- Learn from the past.
- Live in the present.
- Plan for the future.

The first thing that helps me cope is knowing that I control my circumstances and have the power to make changes. I see each difficult situation as a temporary, unique challenge rather than a permanent and insurmountable situation. Second, I focus on progress. I like to say, "I'm not a historian, I'm a futurist." I'm always moving past the past and toward a goal. Which leads me to my third coping mechanism: Don't move away from a problem; instead, move toward a solution. Your history does not determine your destiny; you do.

All problems are asking for solutions; focus on the pattern, what you can learn, and how this problem is just a challenge that will result in a positive outcome. Has something like this happened before? What skill can I practice in resolving it? How can I complete this challenge better positioned than before?

Pandemic

Later in life, as an adult, I used my early learnings and strength to persevere through many other challenges. To survive and even thrive, I have a three-step process.

1) Recognize that the current situation is temporary. At times this is incredibly hard to do as circumstances can be overwhelming. But hardly anything stays the same or lasts forever.
2) Take some time to recognize and process emotions. Although feeling "nothing" helped me survive my younger years, it is not a recipe for a happy life. If this is as challenging for you as it was for me, I recommend reading Brené Brown's recent book *Atlas of the Heart*.
3) Make some progress toward a goal, no matter how small the goal or the progress. Taking action frees us from indecision and doubt.

To illustrate the first step, I am sharing my experience during the pandemic, which seemed like it would go on forever, but we appear to be in a new normal now.

In 2020 I was the chief information officer at Fulton County Schools. The COVID-19 virus became known, and we had several cases in our schools. We formed an executive task force in early January to address concerns and make decisions. By late February, the pandemic was in full force and engulfed our daily lives. We were the first large district to send all students home, and we had to convert the district from in-class to home-based learning.

In that week, my then-fiancé, Scott, discovered significant water damage throughout the first floor and front of my house, and we had to move to his rental home in Canton, Ga. I counted on Scott to handle coordinating the home repair, while I was busy converting the district. What I did not foresee was that, in that same week, Scott's oldest brother would be diagnosed with terminal, untreatable brain cancer.

Any of these events could have brought us to our knees individually, but combined, they were devastating. Each morning, we would give each other a hug, and both declared everything was going to be OK. It took six months, but we were finally able to move back into the house.

We married in Scott's brother's backyard on May 30th so he could share our celebration. We had been thinking of a 300-person wedding, and indeed we did have 300 people attend online, while 10 attended in person.

Scott was able to visit his brother half a dozen times over the next 14 months, and they expressed to each other the love they had shared growing up and together as adults. Despite the use of alternative and emerging therapies, Scott's brother passed peacefully, though he did beat the original prognosis of three to six months.

As we all know now, a vaccine for the pandemic would finally arrive. Even though it changed the world forever, and the people who lived through it will always measure their personal history as before and after the pandemic, today it is little more than a memory for many.

For the second step, I give myself a time limit. For instance, I'll give myself an hour if I have a small emotional issue and no more than three days if it is a big one. Meditation, yoga, or a walk can help me reflect and identify what I am actually feeling and going through. Talking with a friend or therapist can be quite useful at this stage. When I am ready, I will do some physical exercise to change my body chemistry and mood, and then my perspective follows.

Then I move into step three, making lists of action items and checking them off as I move toward my goal. During major job or career changes, I have found that it is important to move toward an exciting new opportunity rather than simply moving away from a less-than-ideal situation. We stay in a better frame of mind when we make life choices that are proactive (empowering) rather than reactive (fear-based).

To illustrate the third step, I'll share this story about beginning to make our dream house a reality during the pandemic!

Growing up and through adulthood, I had always dreamed of building my dream house. I envisioned this later as the place where I could find the aspirational aspect of my name, "serenity": calm, peace, contentment. Little did I know that the man I was dating in 2019 and 2020 shared the same dream. We were talking about homes, and the subject came up. I don't remember who brought it up first. We both had a picture in our mind of what the house would be like, and we both had saved images we had found on the internet. Scott's house was almost identical to the one I had imagined years before.

We decided together that this dream would come true and started working toward it. We started looking at real estate and combing the realty websites, but for months, we found nothing that would meet our needs. We never gave up and were taking a cruise down the river when one of us spotted an overgrown *for sale* sign. We got as close as we could, but the bank was badly deteriorated and several feet above the river. We were still interested. Then we spotted someone on the next dock down reading a book. We drove over to the dock and introduced ourselves. We asked if she wouldn't mind if we tied the boat to her dock and looked at the property next door. She was welcoming and called for her husband. They came back together and showed us around.

The property was rough and overgrown. We learned that it had been on the market for 159 days, and this was during one of the hottest real estate markets in my memory. At first, we thought we should look for an easier spot to build with fewer obstacles. After taking another two or three weeks looking and coming back to this property, we decided to make a bid. We made our bid with three conditions: the land had to be dock-able, the parcel had to be able to support a septic system for a four-bedroom home, and to make that happen, the owner had to have a boundary line survey done and bush-hog the property. Scott cleared the dock-able contingency within two weeks but getting the land bush-hogged and surveyed proved to be difficult during the pandemic. We needed at least three weeks to complete the perk test for the septic, and that couldn't start until the bush-hogging and surveying had been completed.

When we realized we no longer had time to complete these tasks before the closing date of the contract, we asked for an extension. After the property was bush-hogged, it was obvious what a gem we had found, and additional offers started rolling in. Because of this, we almost lost the property, but in the end were able to close.

Empowered

Once we closed, Scott immediately started the process of getting permits for shoreline stabilization and a dock. Scott had to deal with four separate government agencies to get our permits. Getting the shoreline finished and the dock built took almost another two years. We've had a water meter installed and are working on the gas and electricity. We've gone through two sets of plans and several architects. We interviewed several builders before aligning with one. We're getting closer to breaking ground on the house and have completed several other milestones.

The important concept here is that we continue to move forward, even if it is slow-paced. We focus on all the things we have accomplished and that we are building our dream house rather than the turmoil around the most current challenge.

Key Takeaways

- Your history does not determine your destiny, you do.
- Forgiveness brings a kind of peace that allows you to focus on yourself and helps you go on with life.
- Confidence can be earned and learned.
- Do not hold a grudge.
- Regardless of your situation, persevere and stay focused on your goals.
- The current situation is temporary.
- Use negative messages to drive a better outcome.
- Learn from the past. Live in the present. Plan for the future.
- Don't move away from a problem; instead move toward a solution.
- Focus on progress.
- Let go of those who hold you back or tear you down.

Education

A high school graduation at Fulton County Schools.

"Education is for improving the lives of others and for leaving your community and world better than you found it."
~ Marian Wright Edelman

Lifelong Learning

Lifelong learning is essential to staying relevant and active in today's rapidly evolving world. Today, every field is evolving. The skills learned in school, on the job, and throughout our careers must be constantly updated; to stop learning is to fall behind and stagnate. Lifelong learning is essential to the health of our cognitive being. Our brains need ongoing stimulation and new information to continue building neurons. We may forget outdated experiences and information in favor of new learning opportunities. This neuroplasticity enables us to stay current, vibrant, and productive.

In the last few years, we lived through a pandemic unlike anything most people had experienced in their lifetime. The pandemic has changed how we think about work and education, and how we interact and communicate. Being at the epicenter of education during the pandemic, I led the first large district in the United States to turn to remote learning. All eyes in the largest county in Georgia were on our education system to ensure that students continued to learn in this new paradigm, and no one knew how long it would last. We all had to quickly re-learn how to work, live, and go to school in this new environment. It was not easy, and there are long-lasting negative impacts, but we are now in a new era with a shared experience. Some people thought that, after it was over, we would all go back to the way things were before. But history doesn't move backward, and we see now that everything will continue to change.

It is up to each of us to pursue lifelong learning and the confidence that comes with knowledge. We can build a history of winning with small successes: learning to read, reading to learn, mastering the multiplication table, passing algebra. School gives us a chance to learn, grow, and feel successful. With each small success, confidence builds. I thrive in an environment where I can see how my efforts at work make a difference for the team, customers, and organization. Every day, I feel energized when I can help others and contribute.

There will always be people out there who try to tear us down and make us feel not good enough. After reading *Strength Finders 2.0* by Tom Rath, I learned to focus on people's strengths and celebrate their successes as well as my own. I like this approach because everyone has strengths, and this method helps us see them, leverage them, and celebrate them.

Leslie

One of the best examples of lifelong learning is my close friend from high school, Leslie. She is smart and scrappy and has reinvented herself at each stage of life to support her current needs. As a young professional, she started her career as an electrical engineer. Then when she became a mother and wanted more flexibility, she became a massage therapist so she could set her own schedule. She also found her gift to heal people. Later, when her children were nearly grown, her mother was dying, and her marriage was failing. She knew she would need to support herself and pay for her kids' college, so she leaned in on her healing powers and became a nurse practitioner.

Now that her children are adults, she is exploring the world and enjoying her freedom while deepening her healing skills through ongoing workshops and training. While practicing medicine, she has also become a real estate developer in Costa Rica, where she walks on the beach daily and enjoys the catch of the day for dinner. Although our fields of work are very different, both Leslie and I work hard, dream big, balance work with the demands of life, and never stop learning and growing. Leslie is an excellent example of lifelong learning. Every time her situation changed, she pivoted, learned a new skill, went in a new direction, and sometimes started a new career. I applaud her courage, dedication, and commitment to lifelong learning.

It often takes fortitude to keep learning when those around you aren't supportive. The feedback from people who you should be able to trust for support and care can sometimes be devastating. It's important not to let them steal your success, topple your dreams, or believe the reality they try to force on you. People can sometimes be toxic, even those who have a biological relationship with you. These toxic people should be kept at arms-length.

My Future Nearly Stolen

As I mentioned earlier, my father did not pay for my education. It was the last financial lever he had at his disposal to hurt my mother and, subsequently, me. I felt strongly that I needed to go to a four-year college so I could become self-sufficient and fulfill my vision and goal to make a difference in society. I was sure that college was the next step on my path to success. When I received that call from him, it felt like he was yanking the carpet out from under me. I was in shock and cried.

Of course, I immediately told my mother, and she told my stepfather, John. Within a few hours, he rescued my college education. He said, "I have six children and promised each of them that I would pay for their undergraduate college tuition in state. If you go to a New York State school, I will pay your tuition. You will have to work to pay for everything else." To this day, I am grateful that this wonderful person and mentor believed in me. He showed up at this critical junction and helped me in my hour of need. I have been blessed and try to pay this forward every time I get the opportunity.

My mother and stepfather, John, (my "parents") at my first wedding.

By the end of the next week, we had an appointment with the dean of admissions at Stony Brook University. Stony Brook is a top New York State research institution. My mother and I took the two-and-a-half-hour drive. We met with the dean, who looked over my records; my grades and scores were good enough. He said they leave a few spots open for situations like this one, and he admitted me on the spot. He also told me about

Harriman College, where I could get my master's degree in policy analysis and management. My goal had been to get into a college, but now I shifted to getting a master's degree in the same timeframe.

I used my father's words as fuel. I had a chip on my shoulder and would show him that I would not only get one college degree, but I would also have a master's degree! That became my driver: anytime someone said I couldn't do something, or I wasn't good enough, I would use that as fuel to show them I could!

I had an almost perfect GPA my first semester at college. On my report card, I was one A- from a perfect grade and feeling quite proud, until I came home for winter break—my mother took one look at my report card and asked, "What happened?" She was disappointed with the A-, and barely noticed the four As. It felt like I was not good enough for my mother. I didn't feel seen, and she had stolen my joy. I had always felt very connected to my mother and thought maybe her retort was just her way of making a joke. At the time, it was painful because I wanted her to be happy for me. It was not easy to reframe my success and eliminate my mother's judgment. I've always chosen to be resilient and use these moments to push even harder. I wanted to ensure my success was earned and not a fluke of luck.

My father used different messaging than my mother in his judgment of my performance. When I told my father about my grades, he said, "You're not that smart. You must be working too hard and overachieving. Go back and have some fun, and don't try to get A's." His underestimation of my capabilities only fueled my drive to achieve beyond his doubts. (Maybe that's what he intended, but I can't give him that much credit). He had never treated me well or encouraged me to be my best. Although he was incredibly smart and resourceful, a self-made wealthy man who put himself through Cornell in a combined undergrad and vet program, his motives were generally self-centered.

Looking back, I see that he was narcissistic and misogynistic. He did not believe in strong, powerful, successful women. Unfortunately, I ran into far too many people over my career who mimicked that personality profile. The lesson is that we don't have to buy into their reality. We get to write our stories, and I get to choose the story I write.

Carpentry and Homelessness

Between my sophomore and junior years in college, I was thrilled to be able to work on a construction site. It gave me a set of new skills, but more importantly, it opened me up to a new adventure. I started as an apprentice with the carpenter. I felt cool with my ripped t-shirt and carpenter belt. I was a modern cowboy slinging a nail gun. It was a blast. Besides the monetary compensation, I learned a new set of skills I use today with my husband doing DIY projects—roofing, siding, flooring, and drywall. I lived stories that have helped me define myself and that I've remembered all my life. And I got to drive a used red Mustang with a five-speed Hurst gearbox manual transmission!

That summer started my love of sports cars. While most of my friends were working in shops and interacting with other people the same age, I got to work with and learn from skilled tradesmen. I truly believe that this added another important dimension to my life and helped me to understand others. On the job, I met and flirted with the sexy Australian architect doing carpentry. He was charming, and his accent was intoxicating. It was a great summer. Our friendship flowed into my next year at college, and I made trips to New York City to visit him. I loved learning about foreign countries and made it a lifelong goal to travel and understand other cultures. I was raised in rural New York but traveling to New York City gave me a love of cities and the diversity they foster. The two of us explored the city looking for period homes as he measured dimensions. I gained a new understanding of architecture. Years later, I was saddened to learn my Australian friend had perished in the World Trade Center during the 9/11 attack. I found an obituary online with the end of his story, but the experiences and learning we shared will continue with me in my story.

I recently circumnavigated the globe on a business trip and visited three cities in Australia; I couldn't help but remember my friend who had encouraged me to learn about different cultures, develop an appreciation for architecture, and continue my pursuit of lifelong learning.

I wish I could say that my college years were fun and carefree, but that wouldn't be honest or transparent. I worked continually through college to pay for my room and board. During the last two summers before graduation, I was homeless. My mother and stepfather had moved to the other side of the world. I had planned to live with my father, but he was going through his second divorce. My father's estranged wife kicked me out of the unfinished room above the garage where I was "camping" on a mattress. I was on my own. I had to depend on my wits, determination, and the grace of others. I had to fend for myself in every way. I slept on a friend's porch.

My real career started when I was lucky enough to secure an internship at IBM. With one phone interview, I finally felt that my dreams were coming true. I would be working for one of the best companies in the world. Although technology was only a hobby for me at the time, I was thrilled to learn and become more deeply involved in this field.

During this summer, when I was homeless, I had an incredible internship at IBM. It was such a dichotomy! By day, I appeared to be the ideal IBM employee, a model professional. By night, I ironed my only white shirt. I was determined to leverage this opportunity as a bridge to a full-time job. With this mindset I went above and beyond to excel at work despite all the turmoil in my personal life. Each day I went to work with a smile on my face, eager to contribute and learn. This required mental compartmentalization; separating the various parts of my life and focusing only on the current moment and task at hand, a lesson I learned well. I worked hard knowing that I would be on my own from then on. I learned from the past, lived in the present, and planned for the future.

At 22, I graduated with an almost perfect GPA, a BA, an MS, Phi Beta Kappa, a professional resume, and multiple excellent job offers. I now see that my education taught me so much more than policy analysis and management. Through these hardships, I learned to endure and persevere and eventually found success.

Being on my own in college, I sought out mentors to help me push my limits and see other perspectives. I came to know several respected professors and received their guidance. Most successful people can tell you about a teacher who believed in them, someone who changed the trajectory of their life. In high school and college, I was able to find my people, those who did and still do encourage and challenge me to do better.

Christine

At Stony Brook, Christine and I were inseparable; she is still one of my closest friends. We met on our first day of graduate school. It was clear, we'd be friends. Christine, a physics major from Germany, excelled in math, while I held my ground in English.

Post-college, our paths diverged. She entered banking; I ventured into technology sales at IBM. But distance couldn't break our bond. Through divorces and career highs, we cheered each other on. Stony Brook taught us more than textbooks—it shaped our emotional and social skills, vital in the workforce.

Christine embodies lifelong learning. She arrived in the United States as a single mom, pursued physics, earned a master's in policy analysis, and juggled industry shifts. Ski instructor, student of ballet, French chef—she did it all. Today, in addition to her full-time job in marketing at a top credit card company, she and her husband run a ski club and globe-trot. Her lesson? Keep learning, keep growing.

Learning is a lifelong endeavor. I've never stopped reading nonfiction and taking courses to increase my skills. My thirst for knowledge and understanding has always been at the core of my being. The professional certifications I earned earlier in my career—including project management, Enterprise IT Governance, 6 Sigma, and Balanced Scorecard—supported my career progression. I felt they added more resources to my toolkit, more arrows in my quiver. Recently, I've taken several Microsoft courses and certifications to keep up with important technical and professional priorities: accessibility, artificial intelligence, sustainability, and more. Learning new skills is not only important to maintain professional parity; it's also intellectually stimulating. It's important to be well-rounded for executive level conversations.

Tikkun Olam

Tikkun olam is the Hebrew phrase for a central concept in Judaism; to make the world better. I feel particularly drawn to this concept. My focus to fulfill *tikkun olam* is to use my skills, abilities, and strengths to make others' lives better through education. I believe a quality education, like healthcare, and nutrition, should be accessible to all. We can have discourse about "to what level." I advocate for a basic and foundational education that contributes to society. There is a direct relationship with education and the economy of a region. They are inextricably tied together because the improvement of a region's economy improves the quality of life for its residents.

The key to being ready to join the workforce is education. Whether your trajectory is the corporate world, the public sector, or an entrepreneurial endeavor, education provides the foundation and the continuation of a successful career. Education is about acquiring skills. These skills go beyond the textbook and the classroom.[5]

Through education, one can build lifelong relationships, build confidence, learn the importance of diversity and trusting others,

---

[5] Ripley, Amanda. *The Smartest Kids in the World*. Simon & Schuster. 2013

and find inspiration that drives and motivates us. Education helps us build core knowledge and critical thinking skills. Today, there are fantastic alternatives to a four-year institutional education. Online learning and certifications are becoming more commonplace, and going to community college is the best path for many, as it lowers costs and offers opportunities to grow and learn before committing to a four-year degree. There are so many wonderful ways of gaining skills and competencies these days— so when it comes to that, I encourage people to think outside the box if the traditional university/degree route isn't a good fit.

For instance, there are plenty of opportunities with digital, online, and hybrid learning. The use of these tools accelerated during the pandemic, and they are not going away. We are not eliminating teachers. Rather, there is more demand for teachers to provide a wider variety of support, and fewer people attracted to the profession. We need to supplement their tools to reach all students, regardless of their individual challenges, and help them accelerate learning. Digital, online, and hybrid learning environments may enhance the learning experience and increase accessibility, and they are student-centered. This gives teachers and students huge advantages over the old way of teaching through lecturing. It also better prepares students for the real world, which is increasingly technology-oriented.

All of my corporate experience, travel, upbringing, and community work lead me back to education. Transforming education to make it equitable and accessible for all to benefit is my way to practice *tikkun olam*. Education is the great equalizer. Hard work and achievement in education build confidence. Today, as my husband helps me capture my thoughts for this book, I am traveling to Japan to work with the Japanese government to build the future model for education. I've traveled to Europe, India, Singapore, and Australia to understand their educational systems and share knowledge. From these shared experiences around the globe, I've made it my mission to contribute to worldwide education. It is with care, understanding, and empathy that we will move forward to bridge frontiers and challenges in education. No single field affects world outcomes more than education.

AI in Education

Education is not only the foundation to individual success; it is also the gateway to regional economic prosperity and development.

We're all excited (and apprehensive) about the advances and challenges that artificial intelligence (AI) has to offer. AI has been around since the 1950s and has been incorporated into thousands of products and services over the years. Who would have predicted, even six months earlier, that when OpenAI's ChatGPT was released in November of 2022, it would become the most quickly adopted technology in the history of the world? It will change everything, according to experts. Again, I find myself in a leadership role as we navigate the capabilities and challenges of leveraging AI in both K–12 and higher education in every region of the world.

Many educational organizations throughout the world are wrestling with this new technology. Some are harnessing its power for operational productivity, while others are experimenting with new ways to educate and evaluate students. Many are concerned about the long-term implications and the need for guardrails. Those who learn to adapt to our new AI-infused world will succeed. We can only imagine how the world will be in five, 10, 20, and 50 years from now. Regardless of the details, the ability to learn and adapt will remain a focal point of our lives and our society.

Key Takeaways

- Lifelong learning is a must, so always maintain a growth mindset.
- Confidence comes with knowledge.
- Education drives a success mindset; don't let others steal it.
- Learn to endure. Prize resilience and determination.
- Find mentors who help push your limits!

Self-Sufficiency

Self-sufficiency — a value my parents instilled. But too much of it can lead to isolation. My lesson? Learning to ask for help and building interdependent relationships.

Mid-Century Modern

My mother embodied the definition of a can-do attitude. When my mother discovered she was pregnant with me, she and my father were halfway through building their dream home. I was part of the building process. My mother carried me around on her hip as she surveyed the progress and managed the crew for my first six months. My mother stepped up to be the general contractor and managed the project without any training in construction. She found the craftsmen (carpenters, electricians, masons), sourced the components (light switches, hardware, plumbing), and ran the project. As an adult, I have a love for design, enjoy DIY projects, and often tour construction sites. I thank my mother for that inspiration as she gave me an example of being capable and self-sufficient.

My mother and father hired an architect who had studied with Frank Lloyd Wright. He designed our mid-century modern house on the hill and called it "Highlights." It was an upscale modern rambler. The quarter mile-long driveway curved around this masterpiece to the top of the hill to a service entrance and a circular turn-around. My mother rented the steam roller and paved this drive herself. She supervised the building of every aspect of this masterpiece, including the majestic rectangular exotic wood front doors with brass lion-head doorknobs that spiraled upwards to reveal the keyhole. All the living areas were on the main floor.

This house was incredibly special. It represented my parents' creativity and fortitude. It was a physical embodiment of the power of their bond and why they were attracted to each other. Then it became the symbol of their dysfunction as the basement regularly flooded after storms and the grass grew tall from neglect. Later, the cost of its maintenance was at the center of their court battles. As children, we absolutely loved that house. I still have dreams of being there.

My parents were models of self-sufficiency, forgoing the use of a general contractor to save money, and took on this project with incredible foresight and taste for two people still in their 20s. Their attention to detail and planning inspires me today. Seeing the demands of a large family, they built an oversized kitchen (by 1960s standards). There were two islands with hanging glass cabinets above them, a built-in charcoal broil grill, a full-size refrigerator, and a sub-zero freezer. The counter had a built-in motor with attachments for a blender, cake mixer, knife sharpener, and other tools. (This is probably what inspired my love of gadgets and overuse of online shopping!)

The design was efficient and effective, two of the guiding principles I have used throughout life. There were built-in stainless-steel spice racks, paper towels, foil, and wax paper dispensers. They ensured that form follows function, and that spaces were defined by both indoor and outdoor. The entire length of the kitchen looked out over the hill and through the woods. There wasn't another house in sight most of the year, and the sunset streaming in the windows at dinnertime was always a moment to treasure. They knew how to combine style and quality. The cabinets were flat front rosewood, with rosewood dowels for handles. It was all so gorgeous — I hope to create something like it someday in my own dream home.

This house had a built-in vacuum system, which I appreciated since my chores were dusting and vacuuming. You could move the vacuum hose around the house and plug it into a wall socket for activation. All the dust was collected in a bin in the basement that my brother had to empty. There was a laundry chute for us kids to throw our dirty clothes in, and they would land in a bin in the laundry room. The massive rough-cut granite wall in the vaulted living room topped with clerestories was the home's signature. Living in this fashion-forward home gave me an appreciation for architectural beauty. I was soon to learn that even though my parents could work together to build something, this beautiful appearance could be deceiving, and harmony was only surface deep.

As well-equipped as this house was for convenience and beauty, it lacked central air conditioning, and the basement had water intrusion issues, which would become its downfall. I remember being so hot on summer nights that I could not get comfortable enough to sleep. There was an exhaust fan that, in theory, was supposed to draw out the hot air and make it comfortable. Instead, it just made a huge sucking sound that was intimidating and ineffective. I would later build a home that echoed these themes of beauty that helped disguise my soulless marriage.

Empowered

Early Lessons that Led to Self-Sufficiency

When I was seven, we needed a new car, the lawn was overgrown, and we were too poor to afford proper clothing. My parents could not come to an agreement to take care of what we needed, so I decided to take things into my own hands. I gathered the facts, developed my ask—on behalf of my mother—and prepared logical arguments. My mother stood there while I called my father and had an adult discussion with him about the situation and his obligations. I must have been convincing because he did help with some of the things we needed. Not the car; my grandparents gave us theirs instead.

Looking back, it amazes me that my parents engaged me in this discussion and not only let me be in the middle but put me there. Why me and not my older siblings? Why not? I was fierce, passionate, and articulate. At seven years old, I became an adult. Steel is formed in the crucible. I'm no expert in child-rearing, but I do believe that challenging people to do their best is a good thing. Though I don't necessarily agree with my parents using me as a communications platform, it did make me stronger and more confident.

Unfortunately, these situations were a common occurrence.

In fourth grade, I needed glasses, and my father refused to pay the optometrist. Once again, I had to negotiate between the doctor's office and my father's delinquency.

When I was in sixth grade, my brother had orthodontic work, which my father funded. He warned me that if I needed braces, he would not pay. I had a gap in my front teeth that would not be repaired until I was supporting myself. It was another hard lesson in being self-sufficient.

An entry in my junior year high school diary recalled that my father refused to pay for contact lenses, so my mother was taking this to the judge to force him to pay—but first, I was given the task to try to negotiate this with my father. I remember receiving a call from the optometrist because our bill was overdue. I had to call my father and tell him to pay the bill. He argued with me, as I was supposed to pass his arguments back to my mother. These arguments were horrifying, given that my father was at the time a very wealthy man owning several businesses, as well as property, and that paying these expenses were part of the court orders.

He seemed to only deny the expense when it was mine. I feel like it was as if he were saying, "You were never wanted, so why should I pay for you? Take care of it yourself."

It would not be so bad if these were occasional flare-ups where I had to talk with my dad and things would settle down, but that was not the case. After a weekend visit to my grandparents, we came home to find the house had been robbed—antiques, furniture, art from the walls, and even our German shepherd dogs were gone. The following weekend, we had to visit my father, and that's where we found "our things." My father and older brother had broken in and taken what they thought belonged at their house. My mother sued and never recovered any of it. From this situation, I learned that the enemy may be in my own home. I had to be on guard, in a defensive posture, and not trust even those who should be closest. These "lessons" helped me survive in the war-like situation, but they did not serve me well later in life, or in my relationships.

Empowered

During our forced visits, every Sunday evening, my father would sit us down in his office. The leather sofa was soft and smooth. The stereo sat on rosewood shelves and cabinets. The decorations were elaborate and expensive. He would grill us on my mother's activities and try to convince us that she was wrong, not holding up her end of the bargain. He wanted "dirt" on her and would pump us for information. It was tortuous. We would go home in silence and cry to our mother. She would ask us about every detail. Basically, she did the same thing except that she would "coach" us on what to say next time. "Why didn't you say …" or "Next time tell him …" This training certainly made me a better verbal combatant, but it did not teach me about relationship building.

I am convinced that education for young people should include personal finance, critical thinking skills, conflict resolution, and relationship training. It should be part of the curriculum because far too many family homes still lack this expertise.

When I was about 10 years old, I took a photography course through one of my mother's friends. I was enthralled by the technology and magic of creating art through the lens of the camera. I summoned all my courage to ask my father for a camera, telling him about my experience learning how to use one in class. To intentionally snub me and my gender, he said I did not deserve a camera because I was a girl and turned around and gave one to my brother. That hurt. I focused on other pursuits but never forgot his misogyny, although I didn't know that was the name for it. These early lessons taught me to be self-sufficient and that asking someone for something I wanted or needed would never get me to my goal. I had to do it on my own.

I was determined to be self-sufficient because of all the ways I was shown that no one would help me or support me as I grew up. I was with a man for 20 years, yet we led fundamentally separate lives. I thought this was great at the time. I could focus on work and career and didn't have to mess with the drudgery of intimacy or dependence. My mother had given me the example that people who weren't self-sufficient also weren't of any value.

The Dark Side of Self-Sufficiency

Many people of my generation were raised by parents who valued self-sufficiency. It makes sense; our parents lived through the great depression, and being able to support oneself and one's family was their primary objective. Self-sufficiency may be seen as a moral high ground. However, over-done strengths can become liabilities. Self-sufficiency can be a double-edged sword because it is the opposite of collaboration, which is considered a critical skill for success in today's workplace.

My mother was born before World War II. She was the daughter of a holocaust survivor in an era with a common message of self-sufficiency. My parents and my ethnic culture stressed education, hard work, seeing jobs through to the end, and self-sufficiency. This generation frowned upon asking for help and believed in solving problems independently. As children, we got the message that asking for help was a sign of weakness.

My mother brought me up to be self-sufficient. This requirement was communicated both overtly and subtly. Now, I look back on this lesson and believe that, for her, it was aspirational. My mother found herself in an unsupportive marriage to a man who robbed her of her power. She had been a math major at Cornell University. This is not an easy path, even today, and was certainly an incredibly difficult path back in the 1950s. Rather than use her intellectual gifts, she became beholden to my father's wishes, commands, and erratic moods.

My father came from a poor family. He was born in Brooklyn and grew up in the Catskill Mountains. He became a self-made success after going to veterinary school on a scholarship at Cornell, where he met my mother. For spending money, he would work as a handyman around my mother's residence on campus. He had beautiful blue eyes and overflowed with charisma. She fell for this charming man, but he was no Prince Charming, and this was no Cinderella story.

They married when he graduated, and she was only 19. She told me that she knew on her honeymoon that the marriage was a mistake, but they went on to have four children by the time she was 28 years old. My father was physically and emotionally abusive, philandering, callous, controlling, misogynistic, and erratic. Meal menus had to be approved at the beginning of the week, and a separate table in the dining room was set up for the adults to eat after the children ate in the kitchen and whenever he arrived home. My mother wanted to be self-sufficient but instead was in a terrible marriage that turned into thirteen more years of court battles.

My mother's marriage to my father was difficult, and I'm sure that she felt trapped. She was the type of person who wanted to define herself but instead was defined by her husband. She projected all the lessons she had learned onto us. She was suffering for not claiming her power and developing her career after college. Autonomy, financial security, and stability were the things she wanted for us.

From my mother, I was taught that being dependent on anyone else was unbearable. Being alone on a self-sufficient island was the goal. It was only when I was diagnosed with cancer and became dependent on the grace of others that I was finally able to reconcile that being self-sufficient was not all it was cracked up to be. I had to learn how and why to ask for help.

I'm sure my mother thought about leaving my father many times but would not abandon her children and face the social stigma of divorce. She had agreed to take on this "job" in her marriage vows and would try to see it to completion. She did not know how to take the four of us out of this situation to something better. She remained in place until my father ended the marriage by moving in with his girlfriend and taking my oldest brother with him. I was five years old. I remember sitting at the breakfast table, watching my father eat soft boiled eggs on toast in a bowl while he explained that he was getting divorced and that we were all going to be better off. I ran to my room. I cried. I didn't understand exactly what was happening; I just knew it was bad. I knew my life would change forever.

My mother showed great fortitude in this terrible situation. She had finally escaped this abusive marriage. She carried on raising her remaining three children, working, and earning her MBA. When my parents divorced, my mother had to find a full-time job. Previously, she ran my father's animal hospital. She became the director of the laboratory of Northern Westchester Hospital. We had to pick up all the household chores, including dusting, vacuuming, cooking, and cleaning the kitchen. That first year was very difficult. I remember crying every day in first grade for no reason. Sometimes my mother had to pick me up from school because I was so distraught. The other kids teased me and called me a crybaby. In 1969, divorce was far from common, and there was about zero sensitivity to the children afflicted.

My mother inspired me to get a quality education, work hard, defeat the odds, and overcome whatever challenges I encountered. This woman was a force to be reckoned with in her early years. She had thick wavy strawberry blond hair, beautiful hazel eyes, and a body many men desired. Her intelligence, feminine charm, and beautiful smile drew people in, and her fierce spirit and can-do attitude kept them intrigued.

I learned early on that I'd need to fight every step of the way to earn the life I wanted. Nothing would be given to me or come easily. I was powerful and ready for the race—after all, I had survived my horrific younger years. My internal voice would say over and over, "If not me, then who?" I would do whatever it took to be successful on my own terms.

Being self-sufficient means being alone a lot. At times this has its benefits: no one to answer to, no one to judge you, and no one to force compromise. Being alone does not equal being lonely, and even today I still need my alone time to find solitude and reflection. The cost is feeling lonely and isolated when we feel no one can help us, no one cares, no one is there to support and comfort us. It's hard to appreciate your private time unless you have the security that there are people who love, care for, and support you.

Learning to Ask for Help

The idea of interdependence holds immense power for me. It signifies strength in unity, mutual support, collaboration, and the capacity to achieve beyond individual capabilities. However, being interdependent doesn't imply mind-reading abilities. It demands clarity about our goals and needs. Brené Brown's perspective that "clear is kind" resonates with me. Asking for what you want exemplifies this clarity, fostering not just communication but also the chance for mutual support and growth.

Requesting help is a gift not just for yourself but also for the person you're asking. It's a shared joy—everyone finds fulfillment in lending a hand. Life thrives on collaboration. Doing everything alone isolates you from the world, preventing the chance to engage in meaningful collaborations. By shutting others out, you miss out on the invaluable love, care, and support that people can offer. Balancing self-sufficiency with interdependence is crucial. Without the confidence and willingness to seek assistance, life can become a solitary journey.

Before I learned to ask for and accept the help of others, I found myself in situations of deep loneliness because I hadn't reconciled the need for connection with the need for self-sufficiency. I remember being twentysomething in New York City, and feeling poor, alone, and sad. This should have been the time of my life. I was young, beautiful, had a great job, and was quickly climbing the corporate ladder, but I lacked connection. I pursued relationships where my partner wouldn't or couldn't commit and avoided true intimacy.

At the end of my 20s, I married a man who could not be vulnerable and avoided intimacy. We lived together separately, raising children and devoid of true connection. I hadn't put it all together and was wondering what was wrong for decades—why did I feel alone, trapped, like I didn't know who I really was? I was a working mother and there were so many nights I would come home exhausted and alone to take care of my girls. I remember lying on the floor out of emotional and physical exhaustion and letting them climb all over me while I read to them. The marriage dissolved after years of trying to make it work when the lack of connection and intimacy became too much.

I dove back into what was comfortable for me: work and purpose. I took a job in Atlanta. I was the CIO of a large innovative district. I thought that hard work, perseverance, purpose, and self-sufficiency would be enough to fill the void in my life. I worked with over 14,000 employees, yet I felt alone. I had grown up being taught and believing that asking for help would leave me vulnerable and weak. I saw myself at the pinnacle of my career and in the best fitness of my life. It took cancer to open my eyes to reality.

Cancer is a terrible disease that strips us bare, takes away our power, and leaves us feeling helpless. Like so many things in my life, I decided to tackle it head-on. But cancer doesn't let you be the boss of it. I almost gave in. I was almost defeated and let it win. It was because of the love, support, and care of others that cancer didn't win. I allowed my friends in to help me when I was sick. Friends, colleagues, suppliers, peer CIOs, and all levels of people in my organization signed up on the chemo care calendar offering to accompany me to my chemo sessions. I accepted every offer. Each time it gave me the chance to get closer to them and get to know them better. I found a new type of love I hadn't experienced before.

I am forever grateful to my daughters who inspired me to continue fighting, my friends who cared for me, and so many others who met me in my time of need. For the first time, I was forced to ask for help. I learned the joy it gave others to help me. I learned that the love language that I most needed to receive is "acts of service." I finally appreciate others helping me out of love. A part of me had to die from cancer so I could learn to live. I can now accept and feel that close intimate connection when my husband brings me tea in the morning, feeds my cat, and puts up the dishes so the kitchen will look orderly when I come downstairs.

Caregiving

The other message I was given as a young person was that caregiving of others was not valued. Everyone should be self-sufficient, so no one must care for them. Even as children, we were to be seen and not heard. A child's job was to work in school and at home. Childish behavior was not tolerated. When I was upset or hurt, I was sent off on my own to figure it out. I still remember being told, "Stop crying, or I'll give you something to cry about." Any display of weakness was not tolerated; any display of vulnerability was deplorable. It was easy to conclude that everyone should take care of themselves and not worry about anyone else's needs or happiness. Others can fend for themselves, as I did. It wasn't until much later in life that I realized we need other people, and other people need us. Not everyone is able to take care of themselves. Taking care of others is rewarding and fulfilling.

Not long ago, during the holiday season, one of my colleagues, who is super together and prides themselves on self-sufficiency, became overwhelmed with holiday preparation, family, and work. I was enthusiastic when they finally reached out to me for help. Of course, I wanted to help. I don't feel I did much, but it was enough to get them back to being grounded.

Empowered

Having someone else in their corner helped them relieve the stress. I was grateful for being allowed to help and form a stronger bond with the person. They later told me the next time we get together they want to buy me a drink. Accepting gratitude from another person is a way for us to connect, bond, and foster relationships.

Accomplishments and self-sufficiency are rewarding. To learn, only you can "do your homework," but when you're stuck, you need to reach out. It's not about abdicating responsibility; it's about letting other people in and letting them participate in your life and your journey.

My mother showed me how to break female stereotypes. She showed me strength in adversity, and the gumption to do whatever it takes when you're called upon.

The two pictures on the next page are of my mother and me. The first one was taken in 1990 when I was starting out my career. It was at my first housewarming party with friends, colleagues, and my mother and stepfather. I was 25, three years out of graduate school, and just bought my first home in Stamford, Conn. It was a three-story townhouse with two master suites, one of which I rented to a friend and colleague. The party was a celebration of my ability to be self-sufficient, independent, a success on my terms.

To afford a home, I was extremely frugal and made every dollar count. My work clothes were from top business clothing stores like Alcott and Andrews or Brooks Brothers—classic, high-quality professional attire investments to last a lifetime. I rarely went out to eat, choosing instead to eat tofu or peanut butter on rice crackers, sometimes with raisins or chocolate chips. An indulgence was a health club membership. I always had a place to go work out, lift weights, stay fit, and exercise out my stress.

My mother taught me sacrifice. She always stressed hard work, seeing jobs through to the end, and self-sufficiency. What was missing here was compassion, empathy, collaboration, caring, and support.

The second picture is from mid-2022. My mother is 86, in a wheelchair at the memory care center where she has lived for several years now. She is unable to stand or walk and is totally dependent on 24/7 aids for her every need. She barely speaks, but I delight in her reactions when I spark a memory and her eyes light up with her smile. I find joy in giving love and care to a person who does not seem to value these qualities but deeply needs them.

My mother was a beacon of strength. Mom did not believe in soothing or wallowing. The woman who despised dependency is now in a wheelchair, suffering from Alzheimer's, and totally dependent on others for daily survival for everything. I am there for her and take care of everything just like I did for my biological father and my stepfather at the end of their lives. I have no idea how many years she will persist in this state of total dependency. I am grateful that I learned how to care for others before my mother's decline. I had to re-learn so many lessons with a different lens. Independence, courage, and strength were what I needed, along with a good education and hard work to prove my worth.

My mother made me feel like I had to earn love. My mother was wrong. Caring for others is a *mitzvah*, a good deed, and part of a well-lived life. It is not a form of weakness. It is strength. My mother was incapable of soothing, nurturing, and caretaking, but I have learned how important it is for my well-being and for those I love. I thank all the caretakers and nurturers. Caregiving may be difficult at times, but it's why we are here and have family and friends. It gives meaning to the families that we are born into or to the ones we create from our friends and colleagues. Humans are social beings, and we are better adjusted and happier when we care for and love others.

For me, it took getting cancer to learn that being dependent on others did not diminish my value. It allowed others to show their love. I believe it is powerful to be self-sufficient, but I now know it is also powerful to be vulnerable and grateful when people reach out to help us. I am now able to enjoy relationships where I can be both vulnerable and self-sufficient. I can see love in those reaching out to me and enjoy deeper, more satisfying relationships.

Interdependence

Interdependent relationships are ideal because they offer many benefits. Unlike complete independence or unhealthy dependence, interdependence strikes a balance. Here's why it matters:

1. Mutual Support: In interdependent relationships, both parties lean on each other. They share burdens, celebrate victories, and provide emotional scaffolding. It's like having a safety net—knowing you're not alone.

2. Growth and Learning: Interdependence fosters growth. You learn from each other's strengths and weaknesses. It's a classroom where empathy, compromise, and adaptability thrive.

3. Enhanced Well-Being: When you're part of a supportive network, stress lessens. Emotional connections act as buffers against life's storms. You're not just surviving; you're thriving together.

4. Shared Goals: Interdependent relationships align around common goals. Whether it's building a family, launching a project, or navigating life, the journey becomes richer when shared.

5. Increased Resilience: Interdependence provides collective resilience. You bounce back faster, knowing you're not carrying the load solo.

Empowered

So, embrace interdependence—it's not weakness; it's strength woven in connection.

Key Takeaways

- Self-sufficiency is a double-edged sword.
- Asking for help allows the other person to feel needed.
- If not me, then who?
- Accepting gratitude from another person is another way for us to connect.
- Asking for help is not abdicating responsibility; it's letting other people in and letting them participate in your life journey.
- Interdependent relationships, where each person contributes, are ideal.

Purpose

Purpose is a motivator for work and life. Sometimes we choose it and sometimes it is thrust upon us. Your purpose may evolve and change over time as you move through different phases of your life.

Many young people struggle to find their purpose. My suggestion is to reflect deeply but not force it. Life experience will lead you to your calling.

Motivation

"I'm almost 32 years into my journey at Microsoft, and the reason I'm still excited to go to work each day is to help advance our mission to empower every person and organization on the planet to achieve more. It's about connecting with a deeper purpose, aligning what you do with your values, and believing in the impact you can make. When you find meaning in your work, it becomes more than just a job; it becomes a source of fulfillment and inspiration." – Satya Nadella, CEO of Microsoft

My purpose gives me a reason to climb out of bed every morning. I am excited about each workday filled with meetings, tasks, conversations, and content development. It would not be worth it for me just for a paycheck. I resonate with the quote from one of Microsoft's top leaders. I have always felt that it's important to be driven by something larger than myself. My purpose has changed over my career, but it has always been the motivation that drove me forward. My purpose has helped me persevere through the tough times and create goals worth attaining. It continues to give me a sense of direction and help me stay motivated and know that my activities have meaning.

The way I see it, my conception, birth, and survival are only from the grace of G-d. I've felt a duty to find a spiritual need and a purpose for my life. I was born despite my parents' protests. I barely survived my younger years and then almost succumbed to cancer. There is a reason and a purpose for my life: *tikkun olam*, "to repair the world." I practice repairing the world through transforming education with technology and incorporating new technology to empower every learner and educator to achieve more. New technology like ChatGPT can transform education; for me it's energizing and exciting to be able to bring this to education.

I wanted to find out if my views on purpose align with the new world of AI, so I asked ChatGPT — here's what it said:

1. Sense of Direction: Having a clear sense of purpose gives you direction and a focus. It helps you to prioritize and make decisions based on what is most important to you. When you have a purpose, you are less likely to be swayed by external factors and more likely to stay true to yourself.
2. Motivation: Having a purpose can be a great motivator. When you have a clear sense of why you are doing something, it can give you the energy and drive to keep going, even when things get tough.
3. Meaning: Finding your purpose can give your life a sense of meaning and significance. It can help you feel like you

are making a difference in the world and that your life has a purpose beyond just going through the motions.
4. Happiness: Knowing your purpose can also contribute to your overall happiness and well-being. When you are living in alignment with your values and your purpose, you are more likely to feel fulfilled and satisfied with your life.

Overall, finding your purpose is important because it can help you live a more intentional and fulfilling life. It can give you a sense of direction, motivation, meaning, and happiness.

Phase-Aligned Purpose

My purpose has evolved through the phases of my life. During college, my purpose was to show the world and my family that I was more than they had told me I was or could be. This purpose continued but started to change as I had children. My purpose started to shift to building security for them. I wanted them to have the foundation that I never had growing up, and I wanted to give them the childhood I had missed out on. I was motivated to enable my children's gifts, talents, and interests. My older daughter was interested in exploring singing, dancing, and acting. My younger daughter showed interest in robotics, math, and science. I wanted them to have the kind of support I never had growing up. I found programs for them and helped them grow their gifts and talents. I cared for and nurtured my children from infancy, through their toddler years, then childhood.

As they reached adolescence and started the normal process of becoming more independent, I started to feel rudderless. I questioned my core. Who was I? What defined me? What was most important? Perhaps this process of re-evaluation is common as we age and watch our children grow. For me, it felt like there was a gap, a mismatch in who I wanted to be and where I was then. I felt stymied, unfulfilled, and disappointed in myself for not achieving more.

When I started the job of CIO at Florida Virtual School and was in my mid-40s, my marriage and family were becoming less and less my daily drivers, and I knew I needed to adjust my course. It was not in my nature for my ship to remain rudderless. I needed to be sure of who I was. I needed help to get back on course and turned to a professional life coach.

Over six months, I worked with my life coach, exploring my past, current, and aspirational goals. My coach had me examine my life, my early years, where I found joy, and what was still missing. We made long lists, boiled them down, and then tried to put them into the simplest of words. A friend once told me, "If you can't explain something to a five year old, then you really don't understand it." I worked diligently to boil down my life's purpose into the simplest of terms, asking myself, continuously, "Is this simple enough that a five year old would understand it?" At times, I thought the five year old was way ahead of me, but the persistence of my life coach kept me on track. The process brought clarity and focus. Eventually, I was able to identify three "purpose words" to guide my life and was able to explain those words to my inner child: contribution, exhilaration, and serenity.

Contribution

"We rise by lifting others."
~ Robert G. Ingersoll[6]

ORBIE CIO of the year (Inspire CIO), Fulton County Schools. This is in recognition for my contribution to education through technology leadership. I'm surrounded by a few of the people I supported.

Contribution became my "primary driver." It gives me daily direction, motivation, and meaning. During my purpose exploration, I discovered that contribution is embodied in motivation, intrinsic value, and belonging. It materializes when I help organizations and people maximize their potential. This is the "purpose driver" I use to persevere when days are long, or meetings are not as productive as I could have wished.

My life coach encouraged me to re-evaluate my inspirational words from time-to-time, and over the last decade they have been steadfast. At the time I began this quest, I was

---

[6] "Robert G. Ingersoll > Quotes > Quotable Quotes." Goodreads. https://www.goodreads.com/quotes/8119455-we-rise-by-lifting-others

working at Florida Virtual School, the oldest and largest completely online/virtual school serving about 650K students in 60+ countries and every state in the United States. I learned so much there from Julie Young, the founder and CEO, and so many other thought leaders in the organization, such as Pam Birtolo, Cecilia Lopez, and Jodie Marshal. I soaked up information on differentiation, personalization, competency, and mastery from global education thought leaders who supported FLVS and our development, such as Michael Horn, John Bailey, Tom Vanderark, and Arne Duncan. I saw the contribution they were making, and it helped me mold and understand the contribution I wanted to make. We were excited about what we were accomplishing.

I found the atmosphere of contribution exhilarating. The letters and notes from students, parents, and grandparents were heart-warming and instrumental in fueling my passion for leveraging technology to transform education. Some students just needed us for a class or two. Other students took part in full-time studies. We helped enable students who were already pursuing a career like Ally Raisman, the Olympic gold-winning gymnast, to continue their studies. We facilitated homeschooling by finding a learning path for many who opted out of brick-and-mortar schools for physical or mental health reasons. Our rigorous courses, coupled with our highly qualified and motivated teachers, were effective in facilitating the student's journey and connecting with students on a personal level. I learned the importance of enabling each student to learn at their own pace while being supported by a variety of teaching methods and modalities. The contribution was visible every day, and the atmosphere at Florida Virtual was exciting.

I believe that education is the foundation not only for democracy but also for individuals to provide for themselves and their families. This perspective is based on reading and research. In the larger scheme of things, the quality of education improves the strength of the economy. Education and workforce readiness is directly correlated with a state, region, and country's GDP, the primary measure of economic health. Therefore, I believe education is fundamental to an individual's life fulfillment and the greater good of society.

Exhilaration

Scott and I on the Ocoee River in Georgia, living in exhilaration!

Because purpose words are meant to be long-term guiding principles, I needed to examine them and put them into practice before deciding whether they truly fit my life. By focusing on transforming education with technology, I felt that I was already pursuing my primary purpose: contribution.

After identifying my first word for this process, my life coach and I were comfortable that we had identified a foundational piece of who I am and saw how this was ever-present in my life. I was focused on my job and community nonprofit efforts. Now it was time to place the second pillar of my purpose. We pondered together as she guided me to the next step. I wanted something that was attainable yet fleeting so I could strive for it over and over. I craved more joy and happiness, and that feeling, when expanded, became "exhilaration."

I had glimpses of this when my friends and I were in flow, sharing stories, laughing, and enjoying life. I was seeing it in the accomplishments of my work. I wanted more adventure and "Wow!" moments. I wanted this word to extend beyond my work and into my personal life. I wanted a life partner to explore the world, taste the world's cuisines, see great artifacts and museums, experience nature's beauty around the globe, and live life to the fullest. It became clear that I sought more exhilaration, and this would be the next word to guide my life in the future. Finding two great purpose words seemed like a big achievement, and I would have stopped there, had my coach not pushed me to be more well-rounded and stretch my purpose.

Serenity

The third purpose word and pillar needed to really stretch me in ways that would even make me uncomfortable at times. It was by far the hardest of the pillars to find.

I needed something that had eluded my life. I then thought about what was beyond reach — what I sensed was possible for some people, but I was far from achieving. What was truly aspirational and elusive for me was a feeling of peace, safety, calm, and tranquility. My tumultuous younger years spilled into a racing need for educational achievement and professional accomplishments. I craved the adrenaline rush, additional responsibilities, and opportunities to contribute, but this left me little room for quiet, contemplative, and "serene" moments.

Sunset cruise in our boat on Lake Watts Bar, Tenn.

In a moment of clarity, I realized that my actual name was aspirational. As the youngest of four siblings, my mother was hoping I would be easy and quiet, so she named me Serena in the hopes I would live up to my name. I joked that I let her down; I was loud, competitive, and demanded attention. I realized that "serenity" is my goal. It meant the ability to sit quietly, confidently, and feel truly alive. This goal was the most elusive and would take *many* years and many more trials and tribulations. This feeling is now real when we are at our lake house having morning tea and watching the sunrise over the lake.

Empowered

With Scott, I learned how to meditate and be serene. He gets great joy from his moments of peace and wonders why we don't meditate together more often.

Perhaps all three of these life purposes—contribution, exhilaration, and serenity—are never really "done." They are directional and aspirational. These three words feel right. They fit me. They work for now, and for my future, as they have for the previous 10 years. Even decades later, it's clear that I chose well.

These words are like bells. When any of them "ring," I now smile, and my soul delights, as I know that I'm on the right path. A feeling of serenity is more in sight now than it has ever been. I have my perfect partner, and we both love going on adventures. My current job is the pinnacle of my career. It brings together the experiences and learnings from my previous roles. I enjoy traveling to meet customers and partners around the world, fulfilling my early desire to be in a global role.

Scott and I found our dream home site on 350 feet of lakefront where we will break ground on our house soon, which we will call "Serenity." We've named our boat "Serenity and Scotch." Scott likes to tell people I seek serenity and he seeks scotch. We have both found moments of peaceful serenity watching the sunset from the stern of the boat together, silently in the middle of the lake, absorbing nature's grandeur. We love to share these moments with friends and family and hope they find serenity too. Going on our boat or just sitting on the dock or taking a vigorous hike all now feel more peaceful. These are the places and times where my soul delights in serenity.

How to Find your Purpose

For those seeking to find their purpose, I often use this diagram to frame our dialogue. I've seen this diagram multiple times but have never found out who created it so I could give them credit here. The diagram is a simple exercise that anyone can do to help them understand their purpose. When we are in alignment with our purpose, we feel motivated, excited, and full of empowering energy to make progress. Every day may not feel perfect, but the overarching trajectory will take you closer to your goals.

What do you love to do? What gives you energy? What charges your batteries? Time flies when you are doing this activity, and it does not feel like work. List those activities and actions that give you energy versus what leaves you drained. Now detail your list by categorizing each entry with as many areas as it fulfills (You are Paid for it, The World Needs It, You

Love It, and You are Great at It). This is a personal and subjective exercise that requires your own reflection and exploration. There is no right or wrong answer, only what matters to you and the world.

Next, think about your talents and gifts. What comes easily to you? Are there subjects in school or hobbies that you are naturally great at? What things do you do that lead you to receive compliments regularly? Art? Baking? Writing? Fixing engines? Growing plants?

Again, write down all of the things you do well.

Now consider your talents and your passions. Describe how you can use these to contribute to others. Are there opportunities to leverage these gifts to make a difference in your community, or the greater world? If so, you may have found your vocation.

Finally, can you monetize what you love doing? Is there a professional opportunity where you can leverage your talents and passion? If so—and if you have identified how others will benefit from this work—then you have the sweet spot: your purpose!

There's a free set of templates to help with this exercise at this website: 6 Worksheets & Templates to Find Your Ikigai (https://positivepsychology.com/ikigai-worksheets-templates/).

For me, it took many years of learning, working, and soul-searching to find this core. Even now, in every job, I continually look for and fine-tune my purpose. How can I create the largest positive impact possible in THIS role? How can I contribute to others' success? How can I drive business results? How can I build on the work of others? What is my specific purpose in this specific role at this given time?

Serena Sacks-Mandel

Adaptation

The world is always shifting, and technology is constantly advancing. It's exciting (but sometimes exhausting) to always try to be agile, nimble, and flexible. I've found that adaptation has become an overarching super skill. Even in nature, it is not the largest, smartest, strongest, or fiercest that survive. Those creatures that adapt best to change win the evolution race.

To paraphrase Charles Darwin, it is not the strongest of the species that survive, nor the most intelligent, but the one most responsive to change.

As I write this, the age of Generative AI is upon us, and I find myself at the epicenter of this phenomenon. Working for Microsoft, which contributed greatly to OpenAI's monumental success, has been a privilege and huge responsibility. Working in the education industry at the dawn of AI is exhilarating. Education stands to be altered forever, hopefully for the better. Two-and-a-half years ago, I proposed an architecture and prototype solution for leveraging AI to personalize learning for students at all levels. Back then, this seemed to be aspirational. Now, just two years later, it is possible at scale through thousands of platforms and apps. By some measures, the advent of AI has now enabled this aspect of my purpose, transforming education with technology, to be fulfilled. As I consider how to contribute, it is becoming clearer and more evident that this book is my next step. My energy is heightened when I tell my story to inspire others. I am joyful when supporting the success of others. The idea that "true power lies in empowering others" feels incredibly real to me. I feel fulfilled when I lift others up.

Relationships Matter

My family, friends, colleagues, and my husband all help inspire my journey, and I have been able to contribute to their journeys as well. Although my mother taught me to "be an island of self-sufficiency" and not to need anyone, I have found just the opposite to work better in my life. The more I care for and nurture others, the happier and more fulfilling my life becomes.

My network of family, friends, and colleagues nurtures my purpose, and in turn, I contribute to theirs. When I contemplate this network, the term "networking" comes to mind. But it's more than a job-seeking activity. It's about the delight of connecting with like-minded individuals—people who share our passions and walk similar purposeful paths.

Recently, I was given the opportunity to participate in a roundtable for a prestigious worldwide foundation on neurodiversity. They are focused on children's mental health, accessibility, and equity in education. It was an amazing experience to be engaged with thought leaders and royalty on this topic and discuss how we can work together to eliminate the stigma of "different" learners and enable their gifts and talents. There, I met the president of The Jed Foundation (JED), a nonprofit that protects emotional health and prevents suicide for teens and young adults in the United States. This organization supports gamification in learning. A few weeks later, I was asked to support an incredible foundation in London, U.K., that is creating a center to support mental health wellness for children and young adults. We had an incredible conversation about the challenges young people are facing today because of social media, which has led to body dysmorphia and eating disorders in some youth. Then I connected the president of JED with this organization.

I've made it part of my purpose to support organizations that support emotional health, like JED. At work and at school, it is now much more acceptable to not be OK. It's now a normal routine to ask "how are you" and get a real response, not just a generic one like "fine." It's OK to say *I'm tired*, or *burnt out for the week*, or *having a down day*. During pandemic video meetings, many of us were introduced to our coworkers' homes, children, pets, and spouses. We sometimes worked crazy hours. It was common for children to climb on laps and say *hello*, or for a coworker to say *hey, I have to take my child to ___, so I'll continue the conversation from my car.*

In a world that is constantly changing, I've had to modify the object of my contribution, exhilaration, and serenity to maintain a relevant purpose. The COVID pandemic changed so many things. We will forever measure time as before the pandemic and after the pandemic. I am grateful for the ability to work from home — unless meetings and conferences require travel — and for the reduction in the stigma of mental illness. Working from home can be isolating, though. After 20 half-hour meetings in a day, five days in a row, I am fried and often unsure of what was accomplished. I miss some of those hallway conversations. We need face-to-face human interaction to get to know people and connect on a human level. It's fine to share information and even use meetings to create content, documentation, plans, and execute projects, but sitting together and breaking bread at a meal is irreplaceable. We are social beings and need interaction to feel another person's energy and really see them. In this ever-changing world, there is always going to be give and take, and I recognize that video conferencing has been a huge equalizer for height, color, size, and location. More than once, when I've met people face to face for the first time after extended video conference meetings, it has been surprising to see their height—a head on a body, instead of just the face!

When your face is in a box on the screen in a video meeting, differences dissolve. Women are more equal at the table. The politics of human size, height, location, age, and many of our quirks become invisible. If we can't get a word into the dialogue, we can use the chat box, or send an emoticon. Sometimes there's even a back channel where you can talk to an ally about what is happening in the meeting without anyone else knowing. In summary, there are more ways to interact than in person, and so online/hybrid (Teams, Zoom, etc.) meetings are not going away.

Of course, schools are the exception. There were urgent loud demands for students to go back to in-person schools after the initial waves of COVID. Front-line workers need their children to be in school so they could go to work. Office workers may also be relieved to have the house less chaotic while children are outside the home. Teachers and administrators and IT staff were required to go back to school. However, schools had to become more flexible too, and many offer forms of hybrid learning that enable students to continue learning outside of standard school hours.

Although this may seem like a digression from the subject of "purpose," my point is that the world has changed a great deal, and the pace of change is accelerating. Our relationships help keep us grounded and resilient so we can continue to pursue our purpose, even as it may shift. I try to keep in touch with the people I care about through social media, in person gatherings (conferences, events, retreats), and visits with each other. I've also built new relationships by joining associations like those supporting women in technology, and nonprofits such as 1EdTech and The Computer Museum of America, as they are all aligned with various facets of my purpose.

I've found that relationships are essential for our physical, mental, and emotional well-being. They provide us with a sense of purpose and meaning in life. So, it's important to nurture our relationships and invest time and effort into building strong connections with others.

It's not about You

What you may be seeing and hearing when I talk about "purpose" is that our true purpose is not about enriching ourselves. It is not about making more money, buying more things, or being famous. Purpose is about how you contribute to the people around you, how you make them feel, and how you help them succeed. My three words are guideposts for my life, but my purpose, my vocation, my profession was technology sales, organizational leadership, and educational technology.

Now my purpose is becoming coaching, mentoring, inspiring, and lifting people up and paving the way for more women leaders and women in technology. As challenging as it's been for me as a woman in technology, I am well aware that women of color face even steeper challenges. I am an ally and advocate for diversity, equity, and inclusion.

Key Takeaways

- Purpose gives direction, helps motivate, and gives work meaning.

- Purpose helps you live a more intentional and fulfilling life.
- Purpose needs to be long-term guiding principles.
- Find your purpose; then live your values.
- Your purpose is not about you; it's how you CONTRIBUTE to the world and the people around you.

Perseverance

"Appreciate where you are in your journey, even if it's not where you want to be. Every season serves a purpose."
~ The Mindset Journey[7]

Credit for my Three Aunts

Perseverance is the quality of continuing to work toward a goal despite obstacles or setbacks. Sometimes perseverance can be your most valuable tool. I try to maintain long-term focus, set achievable targets, and stay motivated throughout my journey. It is not enough to know your purpose; it must be pursued relentlessly. There have been times in my career and life that were difficult and tiring because of work, travel, or personal commitments. I've needed to push through these patches.

Perseverance has been essential to my career success, because achieving significant goals requires sustained effort and perseverance. Without perseverance, I would have given up too soon, missed valuable opportunities, and failed to learn from my mistakes. The ability to persevere through challenges has been a crucial skill that enabled me to overcome obstacles and achieve my career aspirations. To persevere through those tough days, I

---

[7] www.themindsetjourney.com

focus on adding value, learning, and sticking to my long-term goals.

In my life, I've had plenty of examples of strong, accomplished women—my smart, strong, and accomplished aunts. My mother's three sisters were larger than life; they had gone to top schools, earned master's and doctoral degrees, married extremely well (educated, successful, and devoted), and raised children with ease and aplomb. They all worked hard and broke new ground as women in technology at IBM, the National Institute of Health, and Digital Equipment Corporation. I wanted to stand amongst them, which gave me a reason for my intense drive and perseverance at any cost.

As I studied and grew up in their shadow, I yearned to make them proud. I wanted to make them recognize me as someone who was cut from the same cloth; I wanted to be a Cohen/Hoffman woman like them—my mother's maiden name was "Cohen," and my grandmother's was "Hoffman." A Cohen/Hoffman woman is a winner and does extraordinary things to make the world a better place. I must give tremendous credit to my aunts for helping me overcome some of the greatest challenges in my immediate family and helping form the long-term vision I needed to persevere through my own challenges.

Family Balance

Perseverance helped me sustain myself through challenges at work and unexpected challenges in my personal life. The birth of my first daughter had been relatively uncomplicated. Between my pregnancies with my first and second daughters though, their father decided to leave his stable law practice.

I supported their father's decision because I wanted him to be happy, find his purpose, and be there as a father for our children. It did create some financial instability, but I felt we could manage it together. We had agreed that my career would provide the primary funding in the household. My long-term focus had remained on career development, and if he was home more often, this could serve both our needs.

He took time off and tried out several roles on a part-time basis to see what he liked. It turned out that he enjoyed being a law professor most. I pointed this out to him one evening, and he agreed, so he spoke to the dean about transitioning from adjunct to full-time. The salary and benefits were not as good as he had before, but happiness and purpose were more important. This role provided greater flexibility than his previous position, so he would be available for parenting duties during business hours. I thought we'd still have time as a family during the evenings, weekends, and holidays.

As an additional challenge, late in pregnancy with my second daughter, I was put on bed rest through my due date. She arrived in late February 2000. For me, being idle was a complication to my career and a challenge to me emotionally. At this pre-pandemic stage, when you could not be in the office, there was very little to do.

I was not used to being so idle, and it was difficult not being part of the action. I felt our current home would be too small for us and started looking for something larger. It's dangerous for me to be idle. I shop! My daughter's father had recently made some money in the stock market day trading, and we both thought this was the right way to use it. This was late 1999 and early 2000, during the *dot com* boom. We signed the contract for our new home the day before I went into labor.

We debated on a name for the new arrival as a family. Briana's favorite movie at the time was *The Little Mermaid*, which explains her sister's name, Arielle. Arielle joined our family, with great joy. We didn't foresee the *dot com* bust that happened shortly after her birth, though, which presented another challenge.

Our savings were quickly lost in the bust. We had committed to building and paying for a new house for our growing family, which was finished six months after Arielle was born. A large mortgage had to be negotiated at a high interest rate. We took an interest-only balloon loan. Because of the change in finances, I had to take on a greater share of the household expenses. Their father doubled down and worked more hours to try to make up the loss.

I felt the financial loss was upsetting him greatly and wanted to help him persevere through this setback. I wrote him a letter so he could read it over and over again. It said that I believe in him, and we will get through this together. The money doesn't matter. We can rebuild our nest egg. What mattered was that he was present as a husband and father. I needed the emotional and household support from him so I could focus on my career to financially support the family.

What I had not anticipated was that he would take on the leadership role of the trial team, which kept him away from home many nights, weekends, and holidays. The team practices and competitions usually occur during non-school time. Therefore, in addition to working days and evenings as a professor, he was away on many weekends and holidays. He also took on additional cases and worked relentlessly. It was frustrating to me that he was not helping with the family during these times. It would have been helpful for our family to build positive memories, and I also needed some of this time to prepare for the next day, the next week, or just to recover from career stresses. Before we were married, we had agreed to share in all responsibilities, but he was to have more tasks so I could focus on being the primary wage earner for the family. It was a surprise and stressor to me that I had to both be the primary wage earner for the family and take care of the children by myself on weekends, holidays, and many evenings.

I persevered alone through these years. I was going to "make it," even though I didn't know what that meant yet. I did whatever it took at work and home to make it all happen. This included pumping and storing milk at work for my first daughter for one year and my second daughter for two years. I would book a conference room every three to four hours, cover the windows, put a chair against the door, and face a wall with an electric breast pump for about 20 minutes.

When we had meetings at hotel properties (this happened several times a week), I'd find a green room (normally used for talent to prepare for performances). These turned out to be far easier to use for pumping as they were designed to be private. I laughed to myself at the hoops I had to jump through for my daughters to be healthy while I worked at Disney, "The Happiest Place on Earth."

When I had my children, taking time off was never an option, not only because it is not in my DNA, but because I would have fallen behind the learning and experience curve. My primary focus was career advancement and making money. I understand that, for some people, this trade-off is not reasonable and does not make sense, but for me, letting go of this primary focus on career was never a consideration. We each make our choices based on our personal definition of success, financial needs, interests, and family situation.

At home, thankfully, we had several *wonderful* nannies over the years. They helped fill the gaps—picking the girls up from dance classes after school, making them dinner, and getting them started on homework as they grew. These treasured women often did some grocery shopping, laundered the children's clothes, cleaned up their rooms, and provided endless entertainment in the form of arts and crafts, games, outdoor activities, and much more. I am forever grateful for Carrie, Kara, and Emily and their amazing caregiving for my daughters. They made every day so much easier, especially because I had to persevere when our marriage partnership didn't feel balanced.

We were not in agreement and could not come to a consensus over these years. His priority was to earn back the lost money. I understand his decision. I do not feel it was in the best interest of our family, though. The long hours meant that when he wasn't working, he was sleeping. He was burning the candle at both ends. In truth, both of us were burning the candle at both ends. Neither of us had anything left over to nourish our relationship with each other. I appreciated the money he earned during that period. It helped pay for the girls' bat mitzvahs. It also contributed to their college and wedding funds.

For over a decade we continued to raise our family as our marriage dissolved. I didn't have a healthy model of marriage growing up and thought the only answer was to persevere through the isolation and lack of intimacy.

Having a high-power career and young children is a balancing act. I persevered through these difficult times, and I never lost my long-term focus on career advancement or raising my girls. We had to sacrifice quality family time together. The family time I had needed as a child again eluded me as an adult. Eventually, we began living separate lives in the same house. On weekends, when the Gators played football at home, he was in Gainesville, Fla., with his friends or our daughters. We often took separate vacations. He went to Club Med with male friends, while I spent my time with female friends. I always thought that eventually our marriage would come back together, and things would work out. I did not see the obstacle we had created by not being together at the same time and the damage it was doing to our relationship. We were not spending time together because our interests diverged, and we didn't enjoy each other's company as much as we used to.

I persevered through these difficult times in my marriage and challenges at work. I empathize when I see younger women going through similar struggles today. My focus was on supporting my daughters and being successful at work, so I tackled each obstacle and challenge as they arose. I had been on my own through my childhood and as a young adult. I knew no other way and wouldn't let anything stand in my path.

Career Choice

Perseverance has also been crucial to achieving my purpose in my career. Perseverance has meant staying dedicated to my goals, no matter how difficult they may seem. It requires me to maintain a strong sense of purpose, and continually act toward achieving my objectives. Perseverance has been necessary because success in my career hasn't often been immediate; there have been many setbacks along the way. It took me time, effort, and resilience to achieve my long-term success.

When I was still in graduate school, I interned at IBM in the summer and then interviewed and earned a role at the New York Financial Branch office in New York City. For my last semester of school, I commuted two-and-a-half hours each way, three days a week, and took my classes and research studies on the other days. It was not easy or fun, but I needed this job at IBM to achieve my long-term goals, and a long commute and difficult scheduling would not get in my way.

My first branch role at IBM was looking after our copier business and assisting the sales leader. I found it amazing to go into the skyscrapers in New York City to meet with our financial services clients and help them with their copiers/printers. My role was to read the meter, clear the jams, and explain to the staff how best to load paper and toner. Being allowed in this world and having a purpose was thrilling for me, even if it was to fix a paper jam. It was magical. I could see myself someday playing more important roles. I had planned to live with my boyfriend over summer break, but we broke up before I moved in. My father let me stay in the garage until his estranged wife demanded I leave, which I shared about earlier in this book. I found myself homeless but wouldn't let that deter me. I ended up sleeping on a friend's porch. My long-term goal was more important than my short-term comfort. I remained focused on this future.

I knew I'd be able to support myself and play a role in a big corporation someday, so I was determined to persevere until that happened.

During that semester, I interviewed and tested for an IBM marketing representative role at the IBM branch and was subsequently thrilled to be offered the position. I had also been nominated by Dean Wolf, the dean of Harriman College at Stony Brook where I earned my master's degree, for the Presidential Management Internship program. I hadn't even been aware the program existed or what it entailed because it was so exclusive. The individuals who become finalists earn the opportunity to interview with top federal agencies, including NASA, OMB, HHS, GAO, etc. Only one graduate of each policy analysis and management program can be nominated by their dean. Dean Wolf, whom I deeply respected because of his commitment to doing well by doing good, asked me to apply.

The application process began with a series of essays. Then, a select group would be invited to a full day of interviews in New York City. The interview consisted of individual and group problem-solving exercises and presentations. The finalists were chosen a few weeks later. I was up against top policy analysis and management graduates from Harvard, Stanford, and other top schools. I needed to know that I could compete at this level, and winning the award helped me believe in my future vision. It also forced one of the toughest decisions of my life — whether to pursue a career in the private or public sector.

I couldn't see around the corner of my future. Which would be "better"? What did "better" even mean to me? I was 22 and had experienced so much pain and insecurity in my young life. I had worked in several offices and taken many school courses but didn't know what I didn't know. So, I turned to my wise and intelligent successful stepfather. He pointed out to me that if I first worked in private sector I would be better equipped and provide more value if I later wanted to move to public sector work. This was also a primary theme of my graduate school, and it resonated with me. I took his advice and accepted the IBM marketing representative role. It was prophetic because that was exactly the path my career ended up taking.

My goals had not changed, and the path was becoming clearer. Still, I knew this would be a long journey with many obstacles.

IBM

I started my career at IBM in New York City working with money center banks, international banks, and security firms. When I started, I felt like I knew nothing: I didn't know how to talk to people or how to relate. It was an uphill struggle to learn how the real world worked and operated. I had to re-evaluate every skill I had at the time to strive to do better. I knew this was a great long-term opportunity and wanted to do well. At IBM, I learned sales, management, and professional services and started learning everything there was to know about current technology. Back in the '80s and '90s, this was a new and evolving field, and our technology was the most advanced at the time. Networking, storage, mainframe, mini- and micro-computing, databases, applications, and middleware were all new territory for me to learn. Then came emerging technologies—imaging, workflow, RFID, and eventually the Internet, Web 1.0, then 2.0.

The wonderful thing about technology is that there are no limits; anything that can be imagined can be designed and built, and there is no end to what technology can do. The flipside is that there is a need to continually learn and develop new expertise.

The main obstacle for me at that time was the sheer volume of what I needed to learn. It took perseverance to continually study and learn. Being in technology, this has never changed, and I still spend a large amount of my time learning and understanding the latest technologies. IBM set the groundwork for what I'd need to continue doing for the rest of my career. There are no limits to technology. There is no end to what technology can do. However, this first job taught me to be continually diligent in learning and developing expertise.

The marketing representative role was sales, and sales seemed like the wrong profession for me. I had been trained in a rational decision model: economics, statistics, research, and game theory. I knew how to calculate net present value, internal rate of return, pro forma analysis for return on investment, and total cost of ownership, but I did not know how to establish rapport, uncover requirements, handle objections, and close the sale. I couldn't do the "talk to people" part. I knew these would be big obstacles to overcome. The last thing I wanted to do, I thought, was "SELL." It seemed like a dirty word. I associated it with separating people from their money. I learned about a transactional sell versus a consultative sales process. The latter meant building rapport, understanding problems, requirements, and needs and then finding solutions. Selling became a win/win process where everyone benefited. But that took a lot of time and training. At this point, I realized that perseverance was required to learn this craft to succeed.

I understood that my lack of sales experience was going to be an obstacle to my long-term goals, so I learned consultative selling. I learned how to help people and organizations make decisions and choices that would help them achieve their goals and objectives. During work, I soaked it all in, and after work I listened to tapes of all the success gurus of the day—Tony Robbins, Peter Drucker, Jack Welch, and more. I knew this was a skill I didn't have and needed to obtain.

I had an assigned mentor, who said something profound very early in my tenure—"People buy from people they like." To many, this may seem like quaint, old-fashioned advice, but in 1987, that was gold, and it still is today. It was an *aha* moment; a deep realization occurred for me. It was scary too—I had to ask people how their weekend was, inquire about their children, and create a personal connection. This was a foreign concept to me. At first it was challenging, but this has been a career lesson for me. It's always about the people! In sales, as a teammate, and especially as a leader, it's *always* about the people. For some people this comes naturally, but for me this was a new challenge.

Technological issues can be solved by engaging people with the right set of skills. The more difficult problems to solve are organizational and changing processes and culture. People come first. As a consultant, I saw this repeatedly on assignments to review and assess technical organizations. The CEO would complain that the technology didn't work; the projects were late and over budget; and the internal or external customers were not happy with their services. The problem invariably was ineffective or unclear organizational structure, roles, responsibilities, and expectations. It was always about people.

The best advice for life and work is to ask questions, listen to the answers, and empathize with the other person's experiences. AI can create pictures, answer questions, and can pass a Wharton School MBA test, but it won't make a true connection with people. People want to share their story, be heard and seen. I needed to learn empathy. It took time and perseverance to learn how to connect and relate. Because I experienced little empathy throughout my life and hid behind my survival mechanism (no emotions allowed), feeling empathy for others was not natural. I found Brené Brown's writing and videos helped me understand empathy better.

> "Empathy is a strange and powerful thing. It has no script. There is no right way or wrong way to do it. It's simply listening, holding space, withholding judgment, emotionally connecting, and communicating that incredibly healing message of 'You're not alone.'"
> ~ Brené Brown[8]

She nails it. Reading and learning have helped me grow and continue to provide vision to persevere through the challenges and obstacles, knowing that I can make myself better. Where I struggled with empathy after college, through perseverance, it has now become a strength.

There is always resistance to change, and implementing a new technology or methodology can take time, patience, and perseverance. The things most worth doing are usually the hardest to do. For five years, a large New York City bank was my customer. I was assigned to their retail banking division and worked with a team of experts to design and implement the "Branch of the Future." The idea was to automate some of the manual work that tellers did with ATMs. There was a lot of

---

[8] Brown, Brené. *Daring Greatly: How the Courage to Be Vulnerable Transforms the Way We Live, Love, Parent, and Lead.* Gotham Books: 2012.

resistance to this concept; we heard arguments that people wanted to talk with their teller, and no one would trust a machine over a person. We were sure we had the right idea, and that automation would make banking easier and better. In hindsight, that resistance seems ridiculous because we know what happened: we quickly embraced the convenience of 24/7 ATMs. They have now all but replaced tellers except for the most complex transactions.

I tell this story often to illustrate new technologies that can provide ease of use and reduce costs, but change takes purpose and perseverance. At IBM I began my project management and delivery training. As a marketing representative, it was clear that our customers did not want to buy products; they wanted solutions to their challenges and opportunities. Learning to provide solutions through technology became the next obstacle to my success. I worked across product and services lines to put complete, turnkey solutions together for my customers.

Traditionally, there were experts in technology products and services, but salespeople didn't always understand the business well enough to put together solutions. It required ingenuity and perseverance to find the right people and put together solutions. This was also a change to the way they were currently doing sales, and I had to convince my superiors that this approach would drive additional revenue. Some 30+ years later when I joined Microsoft, I found that, in the education sector, the account executives were primarily selling "products," and when I spoke about providing "solutions," it was exciting and new to many! Everything changes, and yet many things stay the same.

I never saw myself as a salesperson, but I persevered and worked diligently to support my customers and meet my objectives. I didn't have that natural ability, but I could make up for it in effort, and I managed to overshoot my goals every year. After eight years at IBM, I had a big success. I sold a complete white-glove solution to Dean Witter Financial, which was owned by Sears at the time. There was a Dean Witter desk in most Sears stores where the consumer could come in and buy or sell shares of stock on the various exchanges. My project included implementing new data terminals in every location, along with software and cabling services. It was a huge deal that leveraged our hardware, software, and customized operation services. I worked tirelessly, and the project was flawless. I won a significant local award and the Golden Circle, the IBM top sales award.

At the same time, I started working toward a new solution manager role that was being developed. The certifications had to be earned through demonstrated expertise and executive interviews. It was through hard work, late nights, and perseverance that I became the youngest person to achieve this certification. I was then given a double promotion for it. I was 29 years old and getting closer every day to my vision.

I continued to focus on long-term progress and kept accelerating.

Walt Disney World

A year later, I got married and moved to central Florida. At 32, I was eight months pregnant with my first daughter, Briana. My job had me traveling constantly, and I felt I needed something where I could be home more. I interviewed at Walt Disney World (WDW) for an IT management role. They agreed to wait four months for me to join so I could give birth and take a three-month maternity leave. Disney agreed to these terms and made me feel wanted and purposeful in this new role; however, a reorganization and re-prioritization to allow critical focus on Y2K remediation work encroached on my vision and long-term goals to modernize and strategically move the organization forward. As a result, I had to adapt to determine how I could still achieve my goals and objectives within the bounds of these new priorities.

Change is a constant: in almost every organization, in my role, in the organizational structure, the leaders change quite often. The larger the organization, the more often it changes.

From 1997 to 2000, the technology world was focused on ensuring systems did not crash as we turned the clock to the next century. Most systems were programmed for a two-digit year, and the new century would require a four-digit year. There was widespread fear and anxiety over this and an unimaginable number of personal hours dedicated to changing and testing systems. Because we have the benefit of hindsight, we know that the world did not come to an end. Was this because every organization and programmer were so diligent—or because the issue was overstated and fear-mongers won the day? We will never know for sure, however in my humble opinion, far too much time was spent on Y2K instead of delivering new functionality and value to businesses through technology.

I persevered and added value where I could, knowing that, down the road, we'd get back on track.

The job transition from IBM to Disney forced me to learn a new set of skills. The organizational structure and processes were very different. I was hired to be the liaison between technology and finance. I had sold technology at IBM in the finance world but there was a lot to learn being on the client side. I was confident that even though I didn't have experience in this role, my skills were transferable. I wanted to learn everything about Disney and make a positive impact. Being on the client side or buy-side was interesting. Rather than seeing the difference in my past experiences and knowledge as an obstacle, I saw it as an asset. I found that sales skills were important internally and to better understand our suppliers. Negotiation skills learned in my master's program and at IBM have always been useful. The book *Getting to Yes* by Fisher and Ury was a primer for finding win-win solutions to thorny problems when people have varying perspectives.

My focus was career advancement, so I looked for opportunities, challenges, and obstacles to tackle. It surprised me that Disney had many large projects underway without a specific repeatable discipline for delivery. I made it my long-term focus to develop a project management methodology for all the teams to leverage. Using the same processes and tools would ensure delivery and enable the exchange of resources across projects. At that time, the waterfall methodology for project management was in vogue. I learned it in depth, implemented tools, templates, RACI (responsible, accountable, consulted, informed) charts, and trained the staff. In the process, I hit a new obstacle: the director assigned to the operational change management took credit for all the teams' successes, and pointed fingers and blamed others when things didn't go well. I had run into this before, and I knew that adding value and making progress for the entire organization was the way to persevere in this challenge. So, I focused on this goal and ignored the turmoil. In hindsight, if I had let this person know how the team felt and what he could do to boost morale, it would have been helpful to all.

It is important to do great work, collaborate and work well with others, maintain your integrity and authenticity, as well as make rational decisions for yourself and your team. I try to ensure that my contributions have value and are aligned with my organization's goals and view of my role. I also try to check in with my leader regularly to ensure that I am doing the right work and making a difference. Recognition has never been important for me to feel valued, as long as my leader agrees that I am doing well, contributing to the organization's goals, and helping others.

At Disney, my hard work paid off, and I was given the position of manager of financial systems and transactions. For 18 months, every customer purchase and financial transaction at WDW was processed through the systems my team supported. In those days we carried pagers or beepers, and an IT systems manager was paged whenever there was an error in processing, regardless of time. My pager seemed to go off nearly every night between two and four in the morning. I was getting such little sleep with a newborn and these disruptions made me determined to find the root cause and fix it. I worked with the team to develop a root cause analysis process to identify and address the issues underlying the nightly system failures. Quite often, it was due to front-end system failures that then transmitted erroneous information to the financial system. My team took my feedback and developed and initiated an end-to-end testing rhythm for all changes rather than just a stand-alone system test. My long-term work focus was process improvement, which aided my short-term personal goal: to finally get some sleep!

Throughout my career, I have always looked for ways to improve organizational efficiency to stand out and advance my career. I've been successful in finding obstacles and challenges and turning them into opportunities. I realized that a metrics program could help support the objectives and successful outcomes that Disney aspired to, and in addition, risk management could prevent future issues. I also learned that there was an open position to address this gap, so I volunteered to expand my responsibilities and deliver this capability to the organization. I engaged consultants from Norton and Kaplan's Balanced Scorecard organization to mentor me and my team through the development and implementation of a program, first for IT and then several individual operating units.

These stretch assignments, in addition to several others, helped me gain skills, experience, and exposure to new ways of doing things, but they were a lot of extra work. It took additional persistence to finish these assignments. They also brought me closer to others both inside and outside of the organization, and I developed a professional network critical to career development. I learned to look for the role that others shy away from and take it on, like selling to a previously dormant account or learning a new method and teaching others. Success at these activities may bring recognition from leadership, but perhaps more importantly, it builds a portfolio of accomplishments.

Harcourt

In 2002, I was giving a presentation on using the Balanced Scorecard as a tool to develop an IT strategy when I met my future boss from Harcourt. This prestigious firm, Houghton Mifflin Harcourt, traces its origins to the well-known nineteenth-century Boston publishing firm Ticknor & Fields, founded by William Ticknor in 1832. I impressed my future boss with my talk, and he recognized a need that his organization had for my skills and experience. I was hired to establish their program management office. My long-term focus was to transform the organization with a program management process and orchestration. Less than a year after I joined Harcourt, the CIO left the company. As his right-hand person, I had transformed the organization from a stodgy old information system shop to an efficient, effective, and highly functional modern technology group.

We went without a CIO for the next two years. The lack of leadership during this period presented many obstacles and challenges. I ensured that all my managers earned their certifications as professional project managers, which I earned too in 2005. Unfortunately, the new CIO was not in alignment with the on-going transformation. I saw that my career advancement would be curtailed. I had to recognize when I needed to persevere through the current challenges and when it was time to leave. This time I knew the right decision was to leave. I pivoted back to consulting with my trusted and respected colleague Diane Meiller-Cook. We did several engagements together until I found my next landing place as the VP of IT at a hospitality company. I was hired by the CIO, who would become a mentor and a friend.

Hospitality Company

At a large global hospitality company, I was hired as the VP of program management and realized that implementing some basic IT governance and program management principles would improve the overall IT competency. The challenge and opportunity to leverage my recent learnings was beyond exciting! Then within six months I was promoted to SVP, adding strategy, leadership development and quality control to my portfolio.

I identified plenty of challenges and obstacles that could be used as opportunities. The organization was sales-driven and, like most publicly traded companies, short-term focused. The business understood that they needed to advance their technological capabilities, but they didn't know what this would entail. They were struggling to see technology as a strategic tool for executing strategy, which was very common at this time. There was no mechanism for holistically prioritizing and managing IT projects and budgets across the organization. This meant that new priorities were often ad hoc, and projects were not well-planned or executed.

Along with my team, I instituted program management and governance to prioritize and oversee projects, helping the team spot risks and bringing troubled projects back in alignment. I learned every aspect of leading a world-class IT organization from this team. As a leadership team, we strived for a culture of excellence.

We worked vigorously every day to ensure that we were aligned and understood each other's priorities so we could deliver our message consistently to the organization. The challenge was to take the IT organization from being "order takers" to "strategic partners" and to help the organizational leaders understand the power that technology had to deliver on their objectives.

Every day was intense because of the business pressures and demands of the work. Although it took a toll on my personal life, I am thankful for the experience and the opportunity to learn so much. The intensity of the work and our CIO's leadership style brought us closer together. That connectedness and belonging is something I valued then and continue to appreciate because our friendships are strong today as we continue to support each other through life's journey.

This organization, like many others, was deeply impacted by the 2008 economic turndown. We had to right-size the IT organization to meet the new requirements of the business. After having built up a team of IT experts, we now re-evaluated to ensure versatility and success of the new business model. Consequently, difficult decisions had to be made, and as a result, I elected to eliminate my position along with most of my team and move on. I was only there for a few years, but because of great mentorship and teamwork, I garnered knowledge and skills that would propel me to a future CIO role.

Florida Virtual School

My long-term goals had remained career advancement and increased salary. Losing the most lucrative and most prestigious job of my career was a severe blow. I began the process of re-evaluating my long-term goals and values. This new obstacle was different from those I had faced in the past. Having worked long work hours for many years at other companies, I was burned out and missed my children. We traveled to England to visit my sister and her family, and I returned to work as a consultant with Diane Meiller-Cook. We hadn't worked together since 2002. It was a tough environment to restart the consulting business, but I welcomed the flexibility and pace where I could roll my sleeves up and do the work as an individual contributor rather than a leader. I needed this time of unburdening my responsibility for others to recoup, re-energize, and re-evaluate my long-term focus.

I now had time to bike to school with my daughter and pick her up at the end of the day. I went on school field trips with the girls and was spending quality time with them both. I was glad to be more present with my daughters and enjoy these special years. I was helping a friend build a business and getting the personal time I needed. In a way, I see this time as my in-work sabbatical.

It prepared me for my next journey and the beginning of fulfilling my purpose of contribution.

In 2011, through Diane Meiller-Cook's organization (DMA), I started a consulting engagement with Julie Young, the founder and CEO of Florida Virtual School (FLVS), the oldest and largest global K–12 online school. It was governmental, public sector, nonprofit, and a global marketing business for their online content and services. Julie is still a friend, inspiration, and mentor. She continues to win awards and break new ground in education. I am forever in awe of her leadership and vision. At FLVS, I had my aha moment. I had been looking for a more fulfilling long-term direction than career advancement and money. I had gained so much experience over the years, and now I could use it to contribute to the world.

This took me all the way back to that conversation with John, my step-father, after leaving college about first going to the private sector and then taking those lessons to improve the public sector. I thought, "Now I can use these superpowers for GOOD!" My career learnings in leadership and technology could be applied to transform education: "The final frontier for technology."

Until this time, technology in education was relegated primarily to state and federal reporting. It was not seen as part of the instructional practices in the classroom and beyond. I played computer word and math games with my daughters when they were young to help them get a running start, but this would be different. It was a chance to use technology as part of the core curriculum. Technology had previously been forbidden in the classrooms. Computers and cell phones were still the antithesis of learning.

At FLVS, I saw firsthand how we could leverage technology to help students who were struggling academically and/or socially. The unique model enabled students in kindergarten through 12th grade to attend school part-time or full-time from anywhere through the magic of online courses and teachers. Many students who struggled in brick-and-mortar schools thrived in this remote environment. Those who were bullied, felt threatened, had social anxiety, had health issues, and/or neurodiversity often found their haven at FLVS. There, they found students like them in classes and clubs with empathetic teachers who nurtured their unique talents and gifts. Career kids, kids with traveling parents, and kids who were greatly accelerated also found FLVS to be the right fit for their needs.

The CIO was struggling with the role and responsibilities and resigned. DMA was engaged to conduct evaluation of the organizational structure and leadership roles and responsibilities. I worked with Diane Meiller-Cook and Julie Young to define what the organization and this position needed. By the time we had finished defining the role, I knew this was something that I wanted, and we all felt like it was an ideal fit. This leadership team was fantastic. They were innovative, bright, self-reflective, and candid.

We had a wide range of strategic discussions on recruiting educators, filling the top of the marketing funnel, managing our funding sources and formulas, handling external partnerships, building the next generation of digital content, and much more. We worked well together. We freely voiced concerns, provided support to each other, and collaborated as a trusting team. We engaged outside voices such as Michael Horn and Tom VanderArk, the best leaders in this emerging space. We were blazing new paths, working long hours, and loving the challenges.

The FLVS model was a pairing of rigorous online digital content with engaged highly qualified teachers and parental support. Students could enroll anytime and work at their own pace from anywhere while they master the course content. They met with teachers regularly for group and individual sessions, while parents monitored progress and supported their students' learning process. This flexible approach to education was supported by a custom student information system and learning management system. I oversaw the development, enhancements, and deployment of these systems and integration with other cloud-based software, including Microsoft365, ServiceNow, Salesforce, and WorkDay. Back in the early 2010s, we were very advanced with an entire cloud-native technology stack. This was an exciting time in my career. I was no longer persevering; I was thriving.

At home, though, it was a different story. I felt I had found my purpose at work and lost my way in my relationship. I felt alone, unloved, unseen, and unsatisfied. I was persevering to support my children—more on that later.

Fulton County Schools

After divorce, I needed distance to reclaim my life. I knew I wanted to expand what I had learned beyond Florida. Just then, the opportunity to achieve my purpose and help accelerate learning for 100,000 students in Atlanta, Ga., presented itself. I asked for my ex-husband's support as I had supported him through job changes and financial impacts so that he could live his dream. I was energized by the challenges at this new opportunity in Atlanta. However, my perseverance was continually required as I maintained two households and commuted back to Orlando every other weekend. My younger daughter was in high school and I was still "mom-ing" her, doing her laundry, cooking with her, providing a place for her and her friends to socialize and also ensuring there was food in the refrigerator at her father's house. On alternate weekends I often drove to Alabama to see my older daughter in college.

As the CIO at Fulton County Schools, the district surrounding Atlanta, I was positioned to lead the technology direction in one of the largest school districts in the United States. I pursued my passion to develop a model for public school education that could be replicated anywhere for all students. As of the 2012–2013 school year, Fulton had 14,500 full-time employees, including 7,500 teachers and other certified personnel, who worked in 99 schools and 15 administrative and support buildings. Approximately 94,000 students attended classes in 58 elementary schools, 19 middle schools, 19 high schools, and seven charter schools. I immersed myself in learning about the IT team, my peers, and district leadership.

I saw opportunities to improve the organizational structure and led through the lenses of people, process, and technology. One of the first things I did was restructure the IT leadership roles and responsibilities, adding a director of security and centralizing infrastructure under one person. I also formalized the program management office. For all the roles, we created career development plans, including ITIL (Infrastructure Technology Information Library) and PMI (Project Management Institute) training, external conferences for networking and knowledge exchange, and formal talent reviews to identify and nurture high-potential staff for leadership succession planning.

Challenges and opportunities to improve the technical structure were abundant. The education industry was the last technology green field, which was one reason I found it so appealing. I could truly make a difference that would improve students' experience and outcomes.

I created a vision of evolving to a "world class" IT organization out of which we developed a comprehensive plan to execute and achieve it. It began with creating individual development goals for each leader and their teams. We leveraged technology partners and systematically turned it all around in record time. We implemented Six Sigma business methodology for quality improvement that measured how many defects there were in the current process and sought to systematically eliminate them. This resulted in significantly better system uptime, support, and functionality.

Simultaneously, we developed detailed plans for personalized learning. Technology was only one facet of this transformation, which primarily focused on helping teachers adapt through personalized learning simulations, technology training, and change management. There were several repeatable motions for our success, which we shared broadly at industry meetings, including leveraging early-adopter educators to onboard their peers, and establishing student "tech squads" to help set up and onboard other students. The PMO (Program Management Office), project manager, and the lockstep partnership between IT and academics were at the heart of successful implementation. It's always about the people! Technical challenges have straightforward solutions, but the operational change takes time, patience, and perseverance.

In addition to my more-than-full-time role, I set about immersing myself in the Atlanta technology community. We can all meet, network, and speak in our specific communities, both online and in-person. I joined the local chapter of Society for Information Management, the Technology Association of GA, Women in Technology (WIT), TechBridge (focused on eradicating poverty in the community), and InspireCIO. I also created a community for the Atlanta Metro Education CIO/CTOs and directors. I presented at several national educational technology organizations, where I shared our stories and learned from others.

It's so important to know and contribute to others in your community with shared interests and roles. These investments in time pay dividends in friendships, community-building, recruiting, and career development.

The results from several years of program implementation at Fulton showed up in increased graduation rates, improved college entrance test scores, additional students in honors and advanced classes, improved climate scores, and so much more across all demographic groups. Technological improvement was only a part of it, but it was a critical component that led to benefits for educators and learners. I believed in using technology to aid education and had worked hard toward this goal—but little did I know that it was about to be tested in the most adverse of conditions.

At the start of 2020, we began to hear about the COVID-19 virus. I was the CIO of one of the largest, most innovative school districts in the United States: Fulton County Schools. We quickly formed an executive task force, and we met most of every day in a conference room, scanning the news for a local outbreak and creating plans around potential scenarios. Then it happened: an outbreak in one of our schools and then another.

Like we did on inclement weather days, we closed the schools and pivoted to remote learning.

We were the first large district in the United States to make this call. For the most part, we were ready—we had all the resources needed to operate remotely. We had 1:1 devices for students in grades six through 12, digital curriculum, assessment, and one of the most sophisticated data analytics solutions in public education (we won the CIO 100 award). We learned the reality that many students did not have their devices at home and many teachers were not using the digital curriculum. We also had to quickly pivot our security, device management, and network solutions to operate outside of the school network.

It was a crazy six weeks of intense work to get devices in the hands of every student and set them up for success. We worked day and night creating technology hubs across the 90-mile-long district, finding and configuring devices, providing passwords, and handing out devices in a contactless method. We also needed to feed the students who were scheduled for free lunches and get internet access to families that didn't have it. From a technical perspective, most of our network and device systems were set up for on-network control and updates, so we had to overcome that challenge too. Everyone went into overdrive for the first months of 2020, working long hours and making sacrifices of time and energy. We had an *esprit de corps* mentality in the IT group; we all pitched in and did whatever was necessary.

At several schools I visited, the car line was so long that I had to park a mile away and walk to the building to assist the team. Whatever it took, we did what was needed to enable students to continue learning. All of this was contrary to our normal cycle of annual planning, phased implementation, and meticulous communication processes. We had to react quickly to the dynamic of the moment, and no one knew how long the pandemic would last.

I began receiving calls for guidance and assistance from other large school systems across the country. They had learned about our success and wanted to borrow our best practices. We did what we could to help them, but I felt the call to move to a role where I could transition the Fulton model to the rest of the country and maybe the world. When I considered my next step, it came to me organically—my mother was in an assisted living facility for memory care and I was visiting her, wearing my Microsoft jacket. I stopped at the store to pick up some creamer and treats for her, and the Publix employee bagging my purchases, who looked like a high school student, said, "Nice jacket! That would be a good place to work."

Sometimes the universe has a funny way of speaking to us. I've made it a point in my life to listen and respect everyone. You never know where sage advice will be given to you.

Microsoft

I wanted to focus on how my skills and abilities could best be leveraged to support the education industry. I thought my purpose was bigger than my current role. I considered CIO roles in bigger school districts, but they would have required a move. My daughter Briana pushed me to consider roles that kept me in my house near Atlanta. I contacted the VP for U.S. education at Microsoft. I had worked with him and gained his respect as a partnering customer. I always felt it was important to treat vendors, contractors, colleagues, and subordinates with respect. He knew who I was and what I had accomplished.

Over the next few months, I interviewed for and was offered the role of U.S. education customer success leader. As the first person in my role, I had to establish the customer success unit within the education industry to provide proactive and reactive support to enable our customers to better utilize their technology. It was an ideal role for me at the time because I believed in investing in making everyone around me successful. This was the key to my management style, and I continued with that purpose and direction in this job too.

After more than a year in this role, I felt the urge to bring my experience to the global level. When the role of global CTO for education was created, I researched it thoroughly and submitted my application. I felt blessed to be selected as the first education CTO at Microsoft and am grateful every day for the opportunity to serve our customers. Much of my role involves working with sales, engineering, and marketing to ensure alignment and support for our customers' requirements. I meet with the field sales teams and customers throughout the world to share my perspective on current trends and provide advice and guidance on their transformation efforts.

My persistence in finding my purpose and staying true to my vision was now embodied in empowering others to make education available to the world. Perseverance has been my most valuable tool. I've maintained long-term focus, set achievable targets, and stayed motivated throughout my journey. Without perseverance, I would have given up, missed these valuable opportunities, and failed to learn from my mistakes. I've used persistence to overcome obstacles and achieve my career aspirations. I've focused on adding value, learning, sticking to term goals, and empowering others.

> "A strong woman loves, forgives, walks away, let's go, tries again, perseveres … no matter what life throws at her."
> ~ WomenWorking.Com

Key Takeaways

- Perseverance is about continuing to work toward a goal despite obstacles or setbacks.
- Maintain long-term focus, set achievable targets, and stay motivated.
- Focus on adding value, learning, and sticking to long-term goals.
- Implementing change takes patience and perseverance.
- Know when to persevere and know when to move on.
- You may not be able to prepare for every unknown scenario, but when the unexpected happens, basic disciplines and skills can make the difference.
- An empowered team that doesn't hesitate to work together to do the impossible is priceless in a crisis.

Resilience

"We don't develop courage by being happy every day. We develop it by surviving difficult times and challenging adversity."
~Barbara De Angelis

Adversity enters each of our lives. Resilience is our ability to withstand it. Fortunately, resilience is a skill that can be practiced and improved—our ability to bounce back when we get knocked down. Life is stressful, and being resilient doesn't eliminate stress, emotional turmoil, or suffering. Resilience is the ability to work through stress, pain, and suffering, and come through on the other side. The tools I use to be resilient originate in framing the story. What happens to each of us may not be within our control, but how we react to it is something we *can* control. The story we tell ourselves will either propel us forward or hold us back. Regardless of what anyone has said or done to you, only you get to decide who you are and how you show up.

My father was a violent man. He was especially cruel to his second son. He and I were closest in age. We would hide together under the bed when we heard the front door open in the evenings after his golf games or when he returned from the animal hospital. He did not drink alcohol. His rage was emotionally fueled. A few bad tee shots or putts would cause him to fly off the handle. It might be a small thing that triggered an outburst. Emotional damage, especially that inflicted by a parent can be the most damaging and the most insidious. My early childhood was a real-life horror story with a real-life monster. Some of the incidents were so terrible that my brother blocked out the memories until I shared them with him when we were in our 50s. I chose to deal with these traumas and use them as fuel to stand up for myself.

One of these violent memories took place somewhere on the New Jersey Turnpike. About two hours into the drive, my siblings were singing and laughing. My parents should have found this charming and enjoyable, but it triggered my father. He yelled at them to shut up. The warning was soon forgotten, and the noisy interactions resumed. My father suddenly veered off the highway and slammed on the brakes. We were all shaken, but this was only the beginning.

He exited the car, opened the back door, and yanked my six-year-old brother out from the back seat. He screamed at him, hit him, and threw him down the embankment. My father closed the back door, returned to the front seat, put the car in gear, and drove off, leaving my brother crying in the ditch. I could have stayed silent, but instead I chose to plead and beg for my father to stop, go back, and retrieve him.

He finally relinquished, stopped, and roughly retrieved my brother. He deposited him in the back of the car like a sack of refuse. The car was silent for the next four hours of the trip.

At three years old, I decided not to live my life in fear. I would face each monster, even this one. I've chosen to reframe this story as "I helped get my brother back into the car." At three years old I had found courage and became the voice that spoke up to my father's rage. This became a double-edged sword; once my parents divorced, I became the conduit by which my parents communicated their anger to one another. I never once chose to back down from my father. I believe he respected me, but I don't believe he ever liked me. This continued to be my strength and my oppression until I left for college. One can choose to live with courage, or one can choose to live in fear. I make the choice every day to bring my best self to the front, smile, listen, empathize, and solve problems; that's resilience.

I feel the greatest factor behind my success has been my resilience. Resilience is standing up after you've been knocked down. It's bouncing back after a setback. It's having the mental toughness to overcome challenges. I believe that we are not products of our history; instead, we are the result of how we frame what happens to us, and only we get to decide that. A good therapist is effective because he or she helps us gain perspective. They hold up a different kind of mirror and help us reinterpret who we are and who we want to be.

Therapy can be hugely helpful. I was the board chair of the largest behavioral mental health organization in Central Florida for five years. One good thing that came from the pandemic is that more people are comfortable sharing that they are not OK and asking for help. It's cool now, and we all get it. Therapy can help provide an alternative perspective of a story, and it has certainly helped me better understand some of mine.

Resilience is needed most in the crucible moments in life. These are the moments that teach us the most. Success feels wonderful, but we rarely learn much from it. I believe that life is about constant learning and that, in the crucible, we are given the opportunity to learn, understand ourselves, and later reflect.

Empowered

My Biological Father's Death

On December 27, 1998, my biological father died. He had been my tormentor for so many years, and I did not have closure. He was relatively young. We had reconnected after many years of being apart, and he had recently met my daughter. I wanted the father/daughter relationship that had always eluded me and was willing to settle for the grandfather/granddaughter relationship that was never to be. His death left me empty, and I experienced waves of sadness and grief.

After his passing, I went to a medium in the spiritual town of Cassadaga, Fla., to hear from him and sort things out. She was recommended, not just a random "psychic" on the street. I told her nothing about myself or why I was there. I was skeptical, but it was a life-changing moment. The room was not what I had expected. I was thinking of a shadowy place with candles and incense; instead we sat in a normally lit room in comfortable chairs.

The psychic, Hazel, was an elderly woman. I remember thinking she smelled like old people. We exchanged pleasantries. I had expected cards, or a crystal ball, or something else. Instead she just started to speak and tell me what she was feeling. The first thing she felt was my father's one true love and the connection they shared. Hazel said she was hearing his true love's name, saying it sounds like something with an "R" like "Rose." Her name was "Rhoda." Then she said, "He has a different perspective now. He can see how he treated you and your mother." He had deep regrets. He couldn't find balance. He regretted his actions and how he treated others. It wasn't an apology, but it was as close as I would ever get. I could feel his presence in her words.

She didn't use the typical language a father would use for his daughter. She talked about his mercurial nature, his hardheadedness, and his unfaithfulness to my mother. Hazel said he could see it all more clearly through a moral and spiritual lens now and he knew he had not been a good father, husband, or friend to many who had trusted him. She helped me to understand that there may have been a chemical imbalance, and it wasn't all his fault. This moment brought me clarity.

I had already separated who I am and what I do from my father and how he treated me, but these conversations solidified my peace with him and helped with forgiveness. I even see how his treatment of me and denial of support helped fuel my drive to be independently successful. I no longer feel the "I'll show you" energy, but that was my attitude in my 20s. It was the rocket fuel that channeled into learning and work.

The session with the psychic helped me find peace. I understood that his behavior was not my fault, but a result of his own problems. He might have suffered from a mental disorder that was never diagnosed or treated. I felt empathy instead of anger toward him. I have consulted other reputable psychics since then, and they all confirmed that my father had a chemical imbalance or illness and that he regretted his actions and apologized in the afterlife. I realized what he could not say to me when he was alive. I overcame the trauma of his abuse and his death by reframing it in a positive way. With empathy and understanding, I forgave my father for how he treated me. Forgiveness became a powerful path toward letting go of anger and owning my power to manifest my goals.

Layoffs

Resilience has been an important strength in both my personal and professional life. The resilience I learned and practiced by surviving my formative years continues to this day to influence and enable me to be resilient professionally. This is another example of framing myself as the hero of my narrative—the one who overcame—and not the victim. In this way, I positioned myself for future success.

One of the most challenging work situations I dealt with is in this next story.

In December 2008, I had to lay off 75% of my staff. I chose to include myself in this lay-off. I was devastated, thinking about every single person with whom I had worked so hard to make my unit a success. This was the third lay-off in a six-month period. I was nearly in tears as I explained to my staff what was happening. I committed to helping each one of them with whatever was needed to land their next position, whether that was a letter of reference or help with a resume.

The company was providing placement services, but I pledged to go beyond. I wanted them to know I was with them and felt their pain. I knew that having someone you trust believe in you is critical in crucible moments. I had grown so attached to this group of people and all they'd done to help our organization be successful that I wanted to inspire the resilience they'd need to move past this adversity. Everyone who reached out to me ended up in roles that were better suited to their skills. I'm so proud of each one of these people and have maintained relationships with many of them today. I'm proud of their resilience.

The layoffs at Microsoft in 2023 were reminiscent of the 2008 recession. I felt the familiar stab of heartache for my fellow employees who were impacted. Again, I provided connections, references, encouragement, and support to these colleagues. In a conversation with my leader who asked how I was doing, I shared these thoughts.

Those who weren't let go grieve for those who were and long to help them. It's akin to survivors' guilt. Those who remain are concerned about the objectives we still must meet without these valuable resources and their talent. And each person who still has their job is afraid, asking themselves, "Will I be next? Am I doing enough to earn my place on the team?"

At work, we do many things to build collaborative, cohesive, and effective teams. But when economic forces break off parts of the unit, it's jarring and even painful, certainly stressful. I have found that it's our humanity that matters most in these downturns and helps us build our resilience skills. At Microsoft, my leader allowed us time and space to heal and opportunities to express our feelings. Keep in mind that many of those people who leave an organization find opportunities better aligned with their skills and aspirations. So many of the impacted employees demonstrated incredible resilience. Their strength gave me permission to continue to do my work and maintain contact through LinkedIn.

Divorce

My marriage had been unraveling since 2000, but I never let up at work or let my work suffer. I had always ensured that my daughters were a priority, and that our family looked like a model family. I remained focused on what I had to do.

Professionally, I had found the right spot; personally, I was devastated. Fortunately, an in-house coach, Dr. Kathryn Haber, was amazing. Her story is incredible, and she recently published *Fear Less. Love More.* I also engaged an external life coach, Ann Teachworth, who was a true light in my life.

As a couple, we grew further apart. We talked about it. We tried to make changes, but the differences were too fundamental. I was growing more isolated and feeling lonely, unloved, and unseen. I remained persistent, though as we tried counseling and therapists. I tried many therapies to "fix" myself—Body Talk, reflexology, reiki, aromatherapy, neuro-linguistic programming (NLP), cranial-sacral work, and more.

Before this, I had made decisions to leave jobs that didn't fulfill my long-term vision and goals, but until this time, I had not considered making a similar decision in my marriage. I tried to compartmentalize and hadn't considered how much my marriage had affected my career and happiness.

At the time, I didn't understand why I felt trapped, unfulfilled, alone. I blamed myself, as so many women do. I felt I wasn't a good enough mother, wife, worker. I wanted to know how I could do better. I went to the gym, dieted, and meditated. As in my career, I was looking for obstacles and challenges to overcome but wasn't considering that this marriage was the wrong long-term goal.

I had some significant breakthroughs with one practitioner. He asked what I was feeling, and I responded, "I feel like I'm not supposed to be here, as though I don't belong, and I'm not wanted." At this point, I realized I had to make a change. Throughout my marriage, I kept trying to be someone I was not; I was missing out on being my authentic self.

Much later, when I was reading Brené Brown's book *Braving the Wilderness*, it once again struck me how important it is to us humans to "belong" as our authentic selves. The challenge is that these complex, deeply-rooted emotions are like onions—you must deal with multiple layers over time in different ways. It would take another decade or so before I could truly be comfortable as my authentic self in a primary relationship. I was persistent in insisting that my partner needed to love and enjoy me for who I was.

In early 2013, my divorce from my children's father was finalized by a judge. Mixed with the relief I felt, there was also fear of the unknown and resentment. I had trusted my ex-husband to be a true life-partner. I was used to having help, a sounding board, and someone to share life's tedious tasks. For example, I was not good at navigating insurance, making sure the medical statements were accurate, and preparing taxes. Friends who were single encouraged me and assured me that I would be able to handle all of these. It turned out that I could.

However, my biggest fear was the impact that the divorce would have on our daughters. I was afraid history would repeat itself, and they would go through what I had endured as a child. Unfortunately, this fear was realized in many ways.

Finding a place to live was my first challenge. The market was tight, and I didn't know my budget. He had managed our money and made me think everything was hard. I wanted a cheap house to save for my kids. I lost the first bid but found a good deal from a military family. I was scared to get insurance, but my friend said it was easy. She told me to go to Geico.

Empowered

Arielle, me, and Briana (left to right)

The day I moved into the new house, everything changed from fear to elation. I realized that for the first time in a long time, I was free. I could once again be myself! I felt joy and happiness. I had reframed loneliness and fear to freedom.

In hindsight, I wish I'd had the clarity of mind and the "strength of will" to stay out of the drama, listen to my inner voice, and tell my daughters about the divorce in my own way. I knew that was right for them. Today I'm still angry at myself for not being stronger. I realized later that I never fought for myself in the relationship. I never demanded what I needed and wanted in the relationship. I had been treading water, but not swimming toward my personal goals. I had continued in the marriage because of commitment but didn't exhibit resilience or courage—these would come later when I summoned the strength to confront the real issues in myself and the relationship.

I wish somebody had told me what a good marriage looked like. I had no positive role models and just thought that marriage was a commitment to work everything out, running a household, and raising children. I felt that when I said my vows at my wedding, it was forever. My stepfather had told me, "Keep your eyes wide open before the wedding and half-shut after." The message was to just overlook certain things, even if they didn't feel right, because I had made a commitment.

People also talk about expectations. If you have low expectations, you won't be disappointed, but that means you don't go for the home run, which limits your ability to dream big. I learned that lesson from my interactions with my father. He would be hours late or would not show up at all to pick me up for visitation. He consistently let me down, so I learned that if I had low expectations, I could protect myself from disappointment. I took that lesson into my marriage. I had low expectations, and the result was an unfulfilling relationship.

I thought that negotiations and compromises would enable us to have a successful marriage. I wanted to keep things on an even keel and not have arguments. I found out later in life when I *did* have a successful marriage that there was no need to negotiate or compromise because my perfect partner wanted to please me as much as I wanted to please him.

When I was lying flat on my back in an MRI for two solid hours, fighting for my life against cancer, I felt that I had clarity on what I did wrong in my marriage. I took the whole burden on myself, flogging myself for not fighting hard enough for what I needed and wanted in the relationship to make it work.

Scott, the love of my life, helped me understand that you shouldn't have to fight for anything in a relationship. A relationship shouldn't be about fighting for yourself, for your kids, or for space to breathe. A good relationship is self-sustaining and brings joy through the fuel of love. I empathize with people who don't know if they should stay in a rocky marriage or leave. You should always work on it before you give up. There is so much at stake, and divorce is never easy or good.

Ultimately, it was good that I went through the divorce and used my resilience and strength to go on and eventually marry Scott. I finally have a relationship where I feel happy, fulfilled, loved, seen, and supported.

I hope others can learn from my mistakes because there are so few good examples of a good relationship. Neither of my parents provided what I would call a good role model for a relationship, even in their second marriages. I have met many people who did not have good parental marriage role-models either. Perhaps that is why so many relationships and marriages don't work out in the long run.

So, what about the famous statistic that half of all marriages end in divorce? That's true, but only when it comes to first marriages. Second and third marriages actually fail at a far higher rate.[9]

There is hope though. Many of my friends are in fantastic long-term marriages, and others have found deep, lasting love after divorce and remarriage. What I have learned from being in a bad relationship and now a good one is that, when it's right, it's relatively easy. I've learned from multiple bad relationships that I won't always get closure. I have to make my own closure because most people are not capable of having tough conversations.

I have realized, though, what I could have done differently or better.

Because there's no closure from the other person, I've had to create my own closure and my own conclusions and learnings. This is how I bounced back. This is my resilience.

We all make mistakes. We can try to learn from them, but it might take years and years to repair the damage done, as it did with my daughters. I had to be in a different place. They had to grow and mature. I made so many mistakes, and there was an incredible amount of pain.

I've tried to reframe that marriage and learn from it. These are a couple books about healthy relationships that I found helpful:

---

[9] "Revealing Divorce Statistics in 2023." *Forbes Advisor*. https://www.forbes.com/advisor/legal/divorce/divorce-statistics/#sources_section

- *Too Good to Leave, Too Bad to Stay: A Step-by-Step Guide to Help You Decide Whether to Stay In or Get Out of Your Relationship*, by Mira Kirshenbaum
- *The Seven Principles for Making Marriage Work: A Practical Guide from the Country's Foremost Relationship Expert*, by John M. Gottman, Ph.D.

Below is a list that I have found helpful for making a good relationship/marriage, based on my experience and experts' opinions:

- **Connection**: intellectual, spiritual, physical, and emotional. Complimentary skills.
- **Fun**: laughter, joy, excitement, intimacy, shared interests, lifestyle, adventure, vacation list.
- **Values**: honesty, integrity, similar financial goals, complimentary beliefs, acceptance of each other's flaws and quirks.
- **Goals**: support each other's goals and dreams, shared future vision.
- **Communication**: disagree without hurting each other, ability to overcome ruptures, comfort and ease of conversation, shared "origin story" (how we met, when we knew it was right, how we became engaged, etc.).
- **Absence of**: jealousy, control, hostility, substance abuse, stonewalling, contempt, bad hygiene.
- **Nice-to-haves** (makes life easier but not deal-breakers): shared religion, food preferences, political views, life stage, energy level, libido, adventure preferences, bucket-list, music preferences.

Serena Sacks-Mandel

My Stepfather's Death

On May 7, 2017, my stepfather, John, died of kidney failure. He had been the adult in the room when everyone acted like children. He gave me stability, grounding, and a corporate role model. Everyone in life is flawed, but he was always a solid presence in my life and was good to me. I felt honored that, in the end, he called upon me to be with him during these final days, to take care of him and take care of my mother. He chose how to live his life and how to end it by choosing hospice rather than continuous medical treatments.

John was brilliant. He was a Fulbright Scholar, had a Ph.D. in chemical engineering, did postdoctoral work at MIT, and spoke five languages. His favorite saying was, "30 years of school, 30 years of work, and 30 years of retirement." He missed his target by just a few months, passing at 89 years old. For the last 30 years of his life, he grew all his own fruit and vegetables. He became a master gardener, Service Core of Retired Executives (SCORE) leader, and master food preserver, and he taught robotics to middle school students. His students won the state championship year after year. John was a lifelong learner and perfected each skill that he focused on.

John was the father figure I always wanted and needed. He walked me down the aisle, advised me on my career, paid for my college tuition, and was a lifelong mentor. When it came time for me to make a major career decision after college, he was the one who advised me to go to the private sector first and learn how to run an organization, then go to the public sector to show them how to run it like a business. His death was a great loss. He explicitly cut his family out of his will, including me, because he valued education for others. He left a tidy fortune to 4H, Oregon State University, and the University of North Carolina, Ashville.

I want to believe that he had confidence and faith in me. I believe he thought I would not need his money to succeed and that he had already given me the best of him. He taught me how to face challenges, pick the right path, and be resilient. I often took his advice and regretted not having heeded his warnings when it came to marrying. John felt the man I'd pick wouldn't support me in my career. He said there were two types of people in this world: those who can fix things for themselves and those who couldn't. I'm not sure this was the right dividing line. He strongly believed in self-sufficiency, which, as mentioned, has its dark side of independence to the point of isolation.

John knew there were issues with my mother's memory and cognitive ability but never had any tests done or tried to provide for her long-term care. He assigned her money to a trust that was restrictive and managed by a large corporate trust company that makes its money from ensuring trusts are not depleted. John was a strong, powerful man, who conquered many things in life and found many successes. Women are drawn to men like him to find safety and stability. John's death left me with the responsibility for everything he left unfinished. Once again, I had to be the resilient one, recovering from the death of my stepfather to help my mother, who was now dependent on me. At the time I did not know that I would soon be the one needing support and care as I was about to battle the biggest challenge of my life: cancer.

Cancer

My body was my temple. I could always rely on it. I had maintained a regime to ensure my physical and mental health. I had not eaten red meat since I was 17. I didn't use caffeine. I avoided sugar and salty snacks. I worked out four or five times a week—biking, hiking, weightlifting, spin classes, yoga, and CrossFit. I had tried to mentally process my earlier traumas and grief with therapy, faith, positive energy, and the law of attraction. I was vigilant in keeping my mind focused and optimistic. I tried to achieve peak performance by tapping into my energy by using methods from the book *The Power of Full Engagement*. With the multiple traumas I had endured throughout my life, I knew staying fit and healthy physically and emotionally was the key to continued success.

In July 2017, I was in my early 50s. I sensed something was wrong physically. I had started to experience some menopause symptoms, but I had an unusually prolonged period. Some of my girlfriends said that was normal. I looked it up on WebMD, which told me it could be menopause, several other conditions, or cancer. Why is every ailment a potential sign of cancer? I called my healthcare insurance provider's nurse hotline, and she said to go to a doctor. I did not have a regular doctor, so I went to the naturopath practice that was treating my menopause symptoms. She said I had to see a gynecologist that day. I was starting to worry. She spent 45 minutes making phone calls and found me an appointment across town. I went.

They ran several tests and said not to worry, that it was not cancer.

A week later, I received an urgent call from the doctor's office to come in immediately for more tests. I left work and went to the doctor's office. They took more blood and ran more tests and told me it wasn't cancer, but they were not sure of the diagnosis. With a hormone prescription in hand, I went to the pharmacy. As I was leaving, the doctor's office called and said I needed to come back, which I did. Next, they did a pregnancy test. I protested due to my age, but they insisted. It was positive. I was in shock. There was no evidence of a pregnancy from the ultrasound or any other test.

The doctor said I needed to check into the hospital for further tests and to mitigate the risk of internal bleeding. Coincidentally, both of my daughters happened to be in town that Friday for different events. Friends brought them to my hospital room where they serenaded me with their favorite country songs for several hours. It was surreal.

The next day, a doctor came into my room and said it was cancer. Fortunately, my friend Melissa was with me. She could listen, ask questions, and help process the information. I felt that this doctor was insensitive, so I was glad she was with me. My recollection was that the doctor coldly gave me the diagnosis and insisted that all my reproductive organs had to be removed. I refused to accept this conclusion and wanted a second opinion. Agitated, the doctor said, "Call me when it's all over your body."

I wept from the core of my being. I was deeply embarrassed to show my friend this weakness but helpless to hold myself together. I could not bear to be sick and could not face losing an organ, a part of my body. If this happened, I would never be "whole" again and could not live that way.

From my perspective, I couldn't accept cancer. I hadn't been shown any conclusive evidence, just the elimination of other possibilities. I insisted on a diagnostic MRI and being discharged from the hospital. I walked out of the hospital alone, drove myself home, and began to make dinner when the phone rang. It was my gynecologist. She said they saw something on the MRI, and it may explain what is happening. They wanted to proceed with an exploratory procedure.

I called my friend Joan, and she agreed to meet me at the hospital to advocate for me. If I was unconscious during the exploratory procedure, then I wanted someone there who would advocate in my best interest. We were at the hospital from 8pm until 1am on a Saturday night. Unfortunately, when I woke up from the anesthesia, the doctor sadly told me it was not good news. There was still no conclusive diagnosis. I needed to find a gynecologist/oncologist to do more tests. She told me to hurry because the type of cancer suspected was very aggressive and may already have progressed.

It took me the better part of a stressful week to find a doctor who would see me quickly. Fortunately, she was somewhat familiar with the rare type of cancer that was suspected. I met her on Thursday. She examined me and reviewed my charts. Being all too aware of the urgency, she scheduled a complete PET scan for the next day, but it was 90 minutes away. My cousin Vicki, who is calm, rational, and smart, went with me, which helped my anxiety a great deal. The doctor also scheduled a D&C procedure— dilation and curettage, where the doctor removes tissue from inside your uterus—that afternoon for additional diagnosis purposes. It was a miracle that she was able to get the operation scheduled so quickly.

I waited for the results over the weekend, and on the way to a party with a girlfriend Saturday night, the call came. The D&C results were conclusive: it was cancer, extremely rare, and very aggressive. Not the diagnosis anyone wants to receive.

The oncologist explained that they needed to remove my uterus but could save my ovaries. That was slightly better than the other oncologist's plan. I began to come to terms with letting go of my uterus to save my life. Once again, I had to call on what little reserve of resilience I had at the time. I resolved that if surgery would take care of the "problem," then I could move on and get back to my life and purpose. She was scheduling pre-op for Monday and surgery for Tuesday.

Within ten days, I had three surgeries under anesthesia, countless blood draws, and a ridiculous number of other tests. By the time I was in pre-op for the hysterectomy, the doctor and hospital staff knew me well. Each time, I arrived with another friend to support and advocate for me. The morning I arrived for surgery prep at the hospital, the same sweet nurse who prepared me on Friday greeted me. She said I was on her mind on her commute to work. She had no way of knowing I would be her patient again, but we had a connection, and she wished me well. She assured me that my doctor was the best and mentioned that she would be using the DaVinci robot for the surgery. She gave me hope.

I woke up without a uterus. Some people will not understand the extent of this loss, especially since I was done having children. But for me it was losing a part of myself and succumbing to poor health, which equated to weakness, age, and the loss of my identity as a young, healthy, and vibrant woman. How would I ever find my soulmate now? There would be no more dating, no more searching, no more hoping for a happy ending.

I thought that I would feel incomplete and old without all my organs. I was all alone, or so it seemed. My future was cloudy — it didn't matter at that point if I lived to grow old. I was done. My life felt like it was over. I had done what I was here to accomplish, and there was no point in going on.

The cancer diagnosis that I received was so much more than an illness. It was a blow to my self-image, my confidence, my self-worth. I had been single for four years, and dating had been a series of nightmares. When I divorced, I was in the best shape of my life and thought it would be easy to find my life partner, but instead, I was disappointed over and over. The emotional trauma of being told I was pregnant was a body blow. How could that happen at my age? A doctor several years prior had told me that birth control was unnecessary. She actually laughed at me for asking about it.

When I was diagnosed, I had just broken up with the person I was dating for several months. He did not want to define the relationship and be exclusive. I thought he was my last hope. No one would want me now. My self-perception was that I was damaged emotionally and physically, past my prime, and too old to be attractive. (Years later, these were all proved to be untrue).

That afternoon, I was in a hospital room recovering when my friend and colleague, Rocheen, came to see me with a pile of papers to sign. She said she was sure I would want to take care of these today. I knew she wanted to see me and make sure I was OK. She was my angel.

Empowered

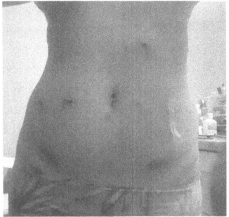

1) 2017 August - Three surgeries in 10 days
2) September - Getting chemo in my cancer sucks T-shirt
3) October - Finalist for CIO of the Year (I won)

The day after the surgery, I went home to recover and nurse my wounds. I was half asleep when my daughter's face appeared. It was not a dream. This 17-year-old took it upon herself to get a plane ticket from Orlando to Atlanta, take an Uber to my house, use the code to get in my garage, and come to my bedside to tell me she loves me and wants me to keep living. Oh my gosh! We had an amazing few days together as she helped me walk around Marietta Square. We ate southern biscuits and went to the farmer's market.

A week after the surgery, I was back in the office for important meetings. My doctor called and said I needed chemotherapy to ensure that the cancer was gone and would not come back. My initial reaction was "No. I'm done." My doctor convinced me to at least think about it. My daughters begged me to do whatever the doctor said needed to be done to beat cancer so I could be there for their graduations and other life moments.

I had so many emotions, but at this point, anger was at the top. I was far too busy to go through chemo. This situation was messing up my schedule. Is this a cruel joke? They had told me it wasn't cancer over and over, but how could I have been pregnant? I was so ashamed. How could I have been looking for my life partner and ended up with cancer instead? This aspect of my illness was a dirty shameful secret. I figured I had done this to myself; it was my fault. I told very few people what type of cancer it was because it was embarrassing.

I wondered, *Why does my life have to be tragic and trauma-ridden?* I refused to own this as my story or learn from this. I would not be one of those people who were changed for the better. I hated cancer and what it was doing to my life. I felt so alone and helpless.

My friend Veronica had been caring for her mother who died of cancer when my stepfather died earlier that year. She came and stayed with me for a while, and we would make fun of cancer and the things people would say that they thought were helpful. For example, when someone would say to me "You are so strong, you've got this," my inner response was, "No, I'm not. I need to borrow your strength." People genuinely cared and wanted to help but too often would ask me what they could do, and I had no idea. Many of my friends who had been through breast cancer knew better; they took care of things that I didn't even know needed to be done.

When the news was out to my family, friends, and colleagues, my living room looked like a florist shop. For some people, flowers bring them joy; but for me, it was sad to watch these beautiful works of nature's art die and then throw them out. It was a sad metaphor. At least the orchids stuck around for a while. I loved and indulged the chocolate. (Thank you!)

Throughout the entire treatment process, I never stopped working. I went from work to the infusion center, then the hospital, and back to work. I never missed a nonprofit or association commitment. Whether it was a surgery infusion scan or other procedure, I just kept going. I was alone at home to flush my line and cry myself to sleep. To survive the struggle, I decided I would rather be at work, with other women, or at meetings, so I just kept going. My team was the best. They helped me continue. They lifted me up in prayer. I prayed for G-d to take me if it was my time or show me how to do this work. Either let me be an instrument for your work or take me. Either way, let me be at rest. Cancer was not the only crucible I survived, but it was the most challenging and the most life changing.

I transitioned from anger to acceptance. In the trajectory of my life, here was another near-death experience, and maybe this was the end for me. I was OK with that outcome. When I reflected on my life, I realized that I had done many important things: I had helped students in my district succeed, I had given birth and raised my children, I had ensured that they could fund their studies. Because I was without a partner, there was no bright future, no retirement plans, nothing that couldn't be done without me. At that point, my bucket list was empty.

I felt that the rest of my life would just be for my own benefit and that I didn't matter to the rest of the world. The odds of me having this cancer were near zero; the fact that I had it did not make sense to me or the medical leader of the American Cancer Institute, who I spoke with. I concluded that life was not just random; it was cruel.

Two months after I was told that the cancer was cured, it returned. I was fortunate to have a family member get me an appointment with one of the two top specialists in the country on this rare type of cancer. My daughter Briana and my sister came with me to the appointment in Boston. The divorce damaged the relationship between Briana and me. The fights we had were awful, and I seriously doubted our relationship and how she felt about me. Prior to going to Boston, I wrote my sister and Briana a letter asking for their support and not their advice. I did not want another round of chemotherapy because there was almost no evidence that chemo would work. I had read there were only four similar cases in the entire country and three had died.

We met with the specialist in the morning. All I knew at that point was that the cancer was back because my numbers were rising again. I spoke to the doctor about what I had read. The doctor at Harvard confirmed that this was very unusual and that there was very little clinical evidence to go by in this case. I was told there were two types of chemotherapy used, one more aggressive than the other. The first step would be to get a CT scan to help determine what was going on, which was scheduled immediately. After the CT, we had a five-hour wait to process the results. It had been a long morning, so we went out for a late lunch.

Later in the afternoon, we spoke to the doctor again. The cancer had metastasized to stage four and had moved to my lungs. Next, it would attack my liver and finally my brain. The cancer was doubling every 10 days. At that rate, I would not live another year. I had resigned myself to my fate. I felt this was all my fault and G-d was punishing me. I deserve this fate for all the damage I had done to my family through the divorce and subsequent move to Atlanta. I had already been through a lot: three surgeries in 10 days and six months of chemotherapy. This time, the cancer had come back even more aggressively. I was looking at an additional six months of chemo, but it would be 10 times the strength it was before and include five additional drugs.

Some of the medications were just to keep me alive, after killing the cancer cells. I felt upset and defeated.

I wanted my daughter's permission to give up and die, but she would not give me permission for that easy out. She figuratively kicked me in the butt. She said, "You have built your life around facing adversity and have built up a storeroom of resilience, now is the time to call on that reserve. You owe it to us, and you owe it to yourself to keep fighting." My sister was clear with the same message, "This is not the time to give up. You are strong and capable." I knew I had to tap into something deeper. I had to focus on the future and step forward. I accepted the challenge because I knew both my daughters wanted to keep me around.

I was now able to shift my mindset and mentally geared up to win this battle. In hindsight, my battle with cancer enabled me to move forward, review my life choices, become more receptive and open, be more empathetic, and learn how to accept love and care from others. It was the beginning of a new spiritual and emotional journey. I also learned that my daughter really did love me and wanted me to live. That evening, we met up with my niece Erika, and the four of us went to dinner. By the time I saw Erika, I had re-gathered my resolve and determination.

At the time, I did not realize how prophetic this dinner would become—they told the restaurant owner the story of why I was there. He refused to let us pay for our meal.

At the school district and throughout the country, people told me that they were giving my name to their congregations, temples, and churches, so entire communities were praying for me. I never prayed for my own healing because so many others were praying for me. It was way more powerful to get prayer from my friends and my loved ones. I prayed that, if G-d wanted me to live, then he should show me what work I have left to do in his name. I saw myself, and still do, as a vessel for G-d's work.

My life mission is *tikkun olam*: to repair the world, help others, and contribute to the happiness of others. When I was in Nepal at 14,000 feet above sea level near Everest base camp, a sign was on the wall that said, "If you contribute to one other person's happiness, you will find the true goal, the true meaning in life." I truly believe that the work I am doing in technology for education is my purpose and why I survived.

Empowered

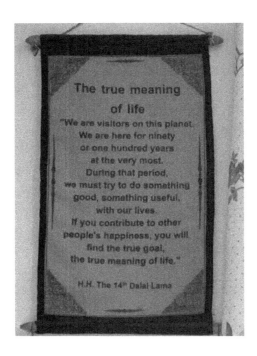

When I was first diagnosed, it was a week before Arielle's senior year in high school started. Arielle was so strong and supportive, and I leaned on her far too much. I still feel bad that she took on an adult role to be there for me and I let her; I was just repeating what my mother had done. My mother had depended on me to be strong when she could not. It cost me my childhood and I regret that damage I've done to Arielle. At the end of the school year, I was cleared of cancer for the second time.

Arielle was selected by her teachers as the baccalaureate speaker at her graduation. She let me know that her speech was about me. Knowing that ahead of time did not stop the waterfall of tears as she described the emotional rollercoaster she went through that year. In the end, she said her strength came from her mother and that she was grateful that I was there to share this moment with her. The crowd erupted in applause and a standing ovation. Maybe it was just me, but I'd like to think I wasn't the only one in that crowd crying.

Being a mother has been my toughest job but my greatest reward. As many mothers feel, I am proud of my children and love them dearly. Without their love and support, I might not have found the will to make the journey. It's an honor and joy to be there for every moment of their lives. And now, I am here for my husband, Scott, whom I never would have met or had this joy and happiness with, if I had not survived.

The lesson here is that, at the end of the day, attitude matters. This is the message I sent to my niece, Erika, who was with me for that dinner in Boston. She began fighting cancer last year at age 26. She remembered my resolve and determination four years earlier. I told her to visualize the chemo as being effective. Thankfully she has recovered after a tough battle.

I'd like to think that G-d saved my life so I could help Erika when it was her time to fight this disease. I know for sure that, for the first six months of my treatment, I was so angry that I didn't care if I lived. I'm sure that this is the reason it didn't work the first time and it did the second time, when I did care.

The second time, I was fighting for the life I envisioned and for my daughters. I envisioned the chemo working. I let people in to help me and accepted their gifts of love and care. I envisioned making a difference in the world in new ways. I kept moving forward and did volunteer work and nonprofit work in addition to my regular job.

I went to a Women in Technology (WIT) event in Atlanta, and somebody asked why I was there with my chemo bag. I said, "I want to be where I can help. I am much happier when I can

Empowered

contribute." One time I went from the hospital directly to a TechBridge event. This is one of my favorite organizations. Their mission is to break the cycle of generational poverty through the innovative use of technology to transform nonprofit and community impact. I was surprised and delighted when offered a ride in a helicopter over the city of Atlanta. It was glorious. It was much healthier for me to be out and about meeting and greeting others and contributing to the world. I learned that if, I wanted to feel better about myself, then I should help someone else. This is why I sponsored an orphan from Nepal on the day I thought my cancer was cured. My victory was to visit her 11 months after the last chemo in Nepal.

Helicopter flight over Atlanta for a TechBridge benefit.

I was not alone. My team let me know this when I entered the building on the first day of work after I lost my hair. My friend and assistant, Rocheen, walked me through every floor of the building. Every woman was wearing a headscarf in solidarity with me, and every man hugged me in prayer. I tried to protest that everyone has illnesses in their family, but they said, "This is for you for what you do for us." It was awesome, and I was speechless. This is what support looks like.

Fulton County administration supporting me through my cancer journey.

I've found the resilience to withstand the adversity that has entered my life. I've established the ability to bounce back when I've been knocked down. Life is stressful, and being resilient doesn't eliminate stress, emotional turmoil, or suffering. Resilience has been my way to work through stress, pain, and suffering, and come through on the other side.

In the end, we are not our history, we are what we decide to be. Some decide more overtly, and others decide by not making a choice, just letting it happen; either way, it's a choice. I choose to decide how I show up, what I do, and who I am.

Key Takeaways

- Resilience is the ability to withstand adversity.
- Frame your story to propel you forward.
- Only you get to decide who you are and how you show up.
- Live your life with courage instead of fear.
- Attitude matters.

Leadership

"Leadership is not about titles, positions, or flowcharts. It is about one life influencing another."[10] This quote by John Maxwell highlights the importance of leadership as a role that is not defined by titles or positions. True leadership is about making a positive impact on the lives of others and inspiring them to achieve their full potential. By leading by example and serving others, you can become a true leader and make a difference in the world. Remember, leadership is not about what you have, but what you give to others!

Leadership and Motherhood

So many books on leadership have been written that it seems inadequate to devote only one section of this book to leadership. It's such an important part of my story that I couldn't ignore or leave it out.

I'll start by saying I've never seen leadership as position- or role-based. It must be earned through trust and taking responsibility. Some of the strongest leaders I met didn't have

---

[10] John C. Maxwell (@TheJohnCMaxwell), "Leadership is not," Twitter, June 3, 2018, https://twitter.com/TheJohnCMaxwell/status/1003244921577508865?lang=en

official titles or roles. They lead by influence, integrity, value, vision, and inspiration. I believe with all my heart that leadership must be focused on developing the individual, giving timely feedback after setting clear expectations to bring out the best in each person. People truly want to follow a great leader. True leadership is about having the mindset and maturity to take ownership of problems, help find creative and effective solutions, and most importantly, care about others.

I've always tried to focus on making a positive impact on those around me. It is always my goal to inspire others to be their best selves, fostering a culture of collaboration and inclusivity, and working toward common goals. Each time I joined an organization, I created a "Culture of Excellence." As an organization and a leadership team, we ensured the performance bar was high and then tried to raise it each year. We would set our priorities on the organization's mission. The goal was focused on creating value in the most efficient way possible and continually improving. We did this by learning to trust each other. As a team, we'd create psychological safety. As a leader, I would model this to my direct reports, and together we'd ensure that everyone was accepted as an individual and that they brought their true self to the group. Rather than "fitting in," I encouraged each person to contribute to the culture. This would make up the unique fabric of our team.

Serena Sacks-Mandel

One of my favorite pictures of my daughters at two and five.

Our careers are inextricably linked to our personal lives. While I was working to build my career, my daughters were young and growing. So many of us have children while we are committed to our careers and leadership roles. When my daughters were little, Briana was a singer, dancer, and actress, and Arielle was a force of nature, smart as a whip, and aggressive too. I have learned so much from these two. It is not easy to raise children. Kids will bring out the best and the worst in you and your partner. It's a long-term commitment... a marathon. If you are committed to being a good parent, you may screw up, but you must always come back to that little person and handle it. They hold you accountable.

Pursuing a career in technology while raising children, alone or with a partner, is not easy. Being a mother has always had a unique set of challenges and obligations. Our babies grow in our bellies, and then we nurse them through our breasts. The physical connection leads to emotional ties. Our hearts are wandering around in the world for all to see. Loving our child is like no other love—it is unconditional and forever.

As our children grow and become more independent, it's simultaneously joyful and painful to witness each year of their growth. I treasured every phase of both my daughters' development, knowing that was time I could never get back. At the same time, I worked fiercely to provide a good life, comfort, and access to resources. Children are born with their own personalities, intuitive knowledge, and interests, and I considered my job to be enabling their exploration and success, but mostly to not get in the way.

For me, being a working mother is inextricably tied to my leadership journey. Motherhood is not required for leadership; it's just how things worked out for me. Whether you are a single parent or married, raising children has its challenges. A friend and colleague whose children were older than mine and already successful once told me that project management is complicated and raising children is complex. Complicated problems may have many moving parts but can be solved systematically by breaking the issues down into smaller parts and then using rules to address each one. Complex problems do not adhere to rules or algorithms. Each situation is unique, and each child is unique. You can take advice from others and your own experience, but you may have to address each situation with an open mind and a problem solver's perspective.

Many mothers say they feel like they are not doing their role as a parent or as a worker as well as they would like. For me, I felt that I was a better technology leader than a full-time mother, so work suited me more comfortably. In the end, for me, work made me a better mother, and being a mother made me a better, more empathetic leader.

Culture of Excellence

Creating a Culture of Excellence requires following a clear and consistent pattern. When I enter an organization, first I get to know the people and understand their strengths. At the same time, I acquaint myself with the organizational drivers and the prioritization of value for each driver. I have worked at organizations that value products over sales, and others that value sales over products. Understanding the prioritization is key to understanding the organization. Next, I set clear priorities that support the organizational drivers. These priorities are reinforced by defining the supporting roles, responsibilities, and expectations and regularly following up on progress. When progress doesn't move in the right direction, I make a point of providing timely, fair, and actionable feedback.

I have joined several organizations that lacked direction and had unclear roles and responsibilities and vague expectations of employees. I found many talented workers who cared about the organization's mission and had the ability, knowledge, or skill to do something successfully but were disheartened by the lack of support and direction from their leadership. Whenever I joined an organization, I made it a practice to meet with each person and get to know them. I wanted to understand who they were, what piqued their interests, their family situation, their hometown, if they moved around a lot, how they chose their line of work, what motivated them to join this organization, and what keeps them there. I enjoy finding a connection with each of my team members.

Empowered

I have found that trust and relationships are best established by finding shared connections or similar experiences. Maybe we travelled to the same country, enjoyed the same type of food, worked at the same company previously, or had children who were similar ages. Whatever connection we could make helps create a bond through our shared human experience.

One of my favorite movies is *Avatar*. An important theme of the movie is that everything is connected spiritually and physically. They emphasize this connection when the indigenous people make a physical connection between them and the animals and plants that surround them. Their bodies are adapted to the surroundings so they can make a physical, emotional, and spiritual connection by plugging into one another. I've tried to view this emotional and physical bonding as part of the process of getting to know people. I wish it was as easy as the connection in the movie, but I've found that, once it happens, we can truly connect and work together.

Using my managing consulting experience, I analyze the organization and determine the optimal structure. Having gotten to know the people and their strengths, I have been able to identify the gaps in the organization, what skills were needed, and how to put each person in a position where their strengths would be featured. Where there were gaps, I often had to either create new positions or re-align the current positions' roles and responsibilities to best meet the organizational needs. Once the organizational structure is established, I focus on creating a Culture of Excellence with the three sides of the triangle: the foundation is Psychological Safety, then Structure & Clarity, and finally deepening our bond through Shared Purpose.

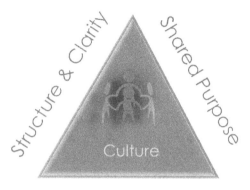

Culture of Excellence

Psychological Safety

The base side of the triangle is Psychological Safety. For a healthy culture, I have found it foundational that each person feels comfortable being their authentic self. It's not enough to talk about diversity and inclusion; it must be demonstrated. When different perspectives and even challenges to the norms are welcomed and celebrated, it enriches the dialogue and helps each person feel seen and appreciated. People need to feel safe enough to try new things and be innovative without a fear of failure. Brené Brown's book *Dare to Lead* has a wealth of information on cultivating psychological safety. *Harvard Business Review*[11] published an article on this topic, and it is now a well-established best practice. Unfortunately, too often, I still hear the practice in many organizations falls far from the standard. I hope that those reading this book take time to learn and practice psychological safety in their organizations.

Trust is essential to an effective team and provides a sense of safety. It is important to build, ensure, and foster trust. Within my teams, we never tolerated sabotaging a teammate or another group. When my team members felt safe with each other, they were comfortable opening up, taking risks, and exposing vulnerabilities. We did this by learning to trust each other.

As a team, we'd also make sure that everyone felt a sense of psychological safety. Too many organizations operate in silos and throw each other under the bus when something does not go well. This toxic behavior undermines the culture and destroys trust. For example, it was never OK for a school technician to blame the security team for the additional sign-in, multi-factor authentication (MFA). Sometimes solutions like MFA were perceived to be less convenient. So, it was important that everyone in the organization understood why security is so important and that the decision to implement MFA was for the benefit of everyone. Everyone needed to be aligned and understand how to "sell" the solution.

---

[11] Gallo, Amy, "What is Psychological Safety?" *Harvard Business Review*, Feb. 15, 2023, https://hbr.org/2023/02/what-is-psychological-safety

Structure & Clarity

Structure & Clarity constitute the next side of the triangle. I provide structure and clarity by establishing clear roles, responsibilities, and expectations for each person, so they know what is required, what success looks like, and how their work supports others. We collectively establish repeatable processes so there are predictable outcomes, and we do not waste time reinventing the wheel for routine activities. However, no one is allowed to use that process as an excuse for why something goes wrong. When challenges and anomalies arise, I support each team member to use their judgment and problem-solving skills to achieve the goal. Empowering team members has been key in every organization that I've run, whether it is customer satisfaction, project delivery, or something else.

I've always empowered and expected the leaders on my team to be leaders in their given roles, to take ownership, and do what is needed for their teams to succeed.

Shared Purpose

The last side of the Culture of Excellence Triangle is establishing a shared purpose. When each person feels an emotional connection to the purpose of the organization, the work becomes (more) meaningful. Like the North Star, shared purpose provides guidance. It forms the basis and direction for decision-making and choices. It also provides a calibration and scale for whether those choices are working toward the shared purpose.

An example of how this works comes from my experience at Florida Virtual School. Julie Young consistently reminded her cabinet that the student should be at the center of every decision, which gave them a Shared Purpose.

When we had a problem to solve, our question was always about how this would affect the student. The solution had to be framed in how helpful it was to the student; then we knew it was the right way to go.

Results

Recently the executive director of infrastructure at Fulton County Schools retired. He invited me to his retirement dinner, although I had already been at Microsoft for over two years. We sat across from each other, and he told me that he was grateful that I had helped him develop as a professional. He trusted me and considered me an excellent leader. I may not have been the smartest or the most technical, but I knew enough to ask the right questions, inspire the right attitudes, and get out of the way to let each person use their expertise to the fullest. As a team, we are highly successful, and because of our collective efforts (with teachers, principals, district operations, and all the other functions), students had higher reading levels, graduation rates, test scores, and school culture scores, as well as more scholarship money. I'm proud to have been able to influence his career and help him be more successful.

Qualities of Leadership

I've built my leadership on qualities I've developed throughout my career and practice every day: influence, integrity, value, vision, and inspiration. Each of these is equally as important as the next, and I try to apply them as a personal measuring stick and compass. When I speak about influence, it evokes the capacity to influence the character, development, or behavior of someone. This is far different from giving commands; it is about setting directions.

Understanding the value that integrity has in leadership is integral to forming trust and respect.

Assessing Leadership

You may be familiar with the 360-degree feedback tool made popular by the amazing coach and author, Marshall Goldsmith. It is an assessment system in which employees receive confidential, anonymous evaluations from the people who work around them. This typically includes the employee's manager, peers, and direct reports. It is an excellent tool for learning how a leader is perceived by coworkers above and below them, as well as peers on the organization chart. I have had them done for me many times and have also recommended this to many team members to improve their effectiveness. In one organization, I had the survey administered, and my self-perception was focused on improving the team's effectiveness and how to deliver projects on time and on budget, while contributing significantly to the organizational strategy.

When the survey results came back, I was surprised not by the overall profile, which was very positive, but by which category had the highest score. Can you guess what it was? I had been focused on communication, problem-solving, and strategic ability. Here I felt that my value was huge, and I was making tremendous positive changes so the overall high score was validating. However, the highest score of them all was "Integrity"! At first, I thought this was a slap: why had they not ranked my strategic ability to lead the team higher than integrity?

My coach explained that having integrity is not a given, not a throw-away answer. I had assumed it *was* a given, thinking, how could a leader *not* have integrity? My coach said that integrity is the foundation for trust, respect, and leadership itself. So, this high score was a testament to my diligence, follow-through, and accountability upon which my team depended. Its importance was a good lesson for me to learn. Integrity and authenticity must come first for a leader; then the hard work of leading can begin.

Leadership Mindsets

In every role, every job, and every day, the first question I ask myself is, "What am I doing to add value today?" Whether I'm in an energizing "flow" moment, or a depleting, challenging time, I try to keep this in mind. The second question I ask myself is, "What can I learn to improve the organization?" With this mindset, I ensure that my focus is on contributing to a positive impact and continuing to grow my skills and experience to ensure the impact is relevant.

I always look for ways to stretch myself and go beyond my current job description to add more. In the late 1990s and early 2000s, I worked for Disney. There was a position as metrics manager open for a long time, and it didn't seem like we could find anyone who wanted to fill it. The organization needed someone for this role, and the gap was becoming apparent. I was the manager of IT strategy, and I wanted to have some stretch assignments to continue adding value and learning. I asked my leadership to give me that role in addition to the one I already had. In this role, I was paired with an outside consultant and learned the Balanced Scorecard. The Balanced Scorecard is a strategic planning and management system that organizations use to focus on strategy and improve performance. The scorecard focuses on strategy and vision. The idea is to establish goals and assume the organization will change behavior and take action to accomplish the goals. It is designed so people will gravitate to the overall vision.

The concepts behind the Balanced Scorecard system flowed with how I envisioned leadership, and I set about instituting it throughout my organization. It added value at every level and helped align people and tasks. Prior to having Balanced Scorecard, the business units were focused solely on profit and loss (P&L). They now have a way to focus on strategy and execution. The process helped create accountability. I've found it to be excellent for adding value to an organization.

Adding value is a message that, as a leader, I try to carry to every person in the organization. I try to say, "Let's create excitement, not confusion." Change not only needs to add value, but it also must be communicated well. We need to keep the customers informed and train them in new technology. When I was at the Fulton School District, we rolled out a new telephone system. We were excited because we believed we were enabling the schools with new and better technology. We had failed to communicate the change and the value of the change, though, so the building leader was surprised by the change when he arrived that morning. This was a learning moment for me and my organization. Since then, I've ensured that we always focus on operational change management, as well as technology change management. The team learned and never let that happen again.

Visionary Leadership

Visionary leaders are highly valued and can transform organizations. Visionary leaders invigorate and motivate an organization. They look to the future and see what they could be. Change is often difficult, and moving toward the unknown can be daunting. It's important for leaders to be visionary and have a natural knack for seeing what is possible. This can be challenging because the leader has a vision, and not everyone will see it as quickly or as clearly. Each person has their own process of absorbing change. It's up to the leader to convey the vision so the organization can execute it.

At times, I felt I clearly knew what was needed, while others needed more time to analyze, assess, and map out the steps. The skill of understanding what "could be" (envisioning) has always been something I've tried to leverage as a strength. My vision at Fulton helped my team see the possible and inspired them to achieve the world-class ecosystem and customer service I described to them so they could make it materialize.

> "Take the first step in faith. You don't have to see the whole staircase, just take the first step."
> ~ Martin Luther King, Jr.[12]

During most of my career, my roles included working with emerging technologies. One of my top strengths in the Strength Finders survey (a survey in the book I mentioned earlier) is "futurist." I am energized by the potential for a better future—for technology to help us live more comfortably and work more

---

[12] Marable, Manning, and Leith Mullings. *Let Nobody Turn Us Around: Voices of Resistance, Reform, and Renewal: An African American Anthology.* Lanham [Md.]: Rowman & Littlefield, 2000.

effectively. Here's one example: If you have stayed at a Disney Resort in the last 10 years, you were probably given a bracelet for each family member that served as your room key, tickets, and on-property "wallet." This bracelet is not only convenient, but it also drives sales. Users can use it as cash anywhere in the parks: they wear it to go swimming in a pool or waterpark, to buy concessions or souvenirs, or to go out to dinner. They can leave their wallets and purses behind. It can also be used by parents to track their children, give them money, and not worry about losing a separate room key or losing your tickets.

Today, this may seem commonplace because we are used to automation like tapping our credit cards and even using cell phones to make payments; generally, we have limited our use of hard currency. This type of technology was almost unimaginable back in the late '90s. From my experience at IBM supporting the finance industry working with devices that either actively or passively give off a unique radio frequency, I pitched the idea to our leadership at Disney. I worked with suppliers, developed prototypes, and then cost-justified it all the way up to the senior vice president of finance.

It took three attempts before it became a reality. The first time it stalled was due to the threat of Y2K shutting down all systems. The second time, during the 911 attack, it was halted again. It took another 10 years for it to be championed and finally brought to the market, and by then I had left Disney. I went back to Disney World on business trips and kept a couple of these bracelets on my "museum" shelf in my office, next to my Galaxy Android 3, Zune, Palm Pilot, Blackberry, and Casio Calculator!

Empowered

My 1970s Casio CQ-1, Calculator, Stopwatch, Clock, Calendar, and Alarm Clock

Be Inspiring

One of the most important duties of a leader is to inspire their people. I believe in creating a vision of excellence, where each person understands how they are contributing to delivering the goals of the organization and how they bring value. As a leader of people in the CIO role, I set out to support individuals; make the organization functional, effective, and efficient; and deliver value to the organization. To achieve these goals, I had to inspire those in my organization to give their best every day.

When I unpack how I inspired my team to be awesome, it took decades of hard work—pressure tested by personal challenges—and my relentless quest for learning and improving. I brought a vision of excellence to them, coupled with confidence in their abilities and support for their personal and professional challenges. My team would tell you that I was approachable, authentic, always had an open door, and welcomed a chat. I tried to take 30 minutes to walk the floor multiple times a week and say "hello" to individuals, find out how their family was doing, and learn more about what they were working on. I learned so much about the work that was being done and the people themselves from these informal chats.

We also had celebrations. At the district, we celebrated the end of the school year and end of the calendar year. Sometimes we did potluck lunches, and other times suppliers would provide food. We decorated on a shoestring budget, and we recognized and celebrated top performers for each group. So, I didn't just tell them about the culture we were building, I modeled it through my actions and interactions.

I found out from my team that I inspired them. Throughout my tenure, people would drop things off when I was at meetings for me to find later. I was touched by the notes of genuine appreciation, a small statue with the words "You inspire me," and the celebrations on "Boss Day" and my birthday. One of my direct reports told me they drove 90 minutes each way to the office because of my leadership—he said it was worth it. Over the years, as my colleagues and I have changed positions and companies, we have still kept in touch, and I continue to support their careers with coaching and references.

Value in Feedback

In my career, I've found that one of the hardest and most important parts of being a leader is giving feedback. Everyone screws up from time to time. Feedback is one of the hardest things to give and receive. I have been effective by building trust first, and I have a very specific way of giving feedback. When you have a team that laughs together and trusts one another, you can accomplish anything. Feedback is only effective when there is trust first. My favorite book on this topic is *Thanks for the Feedback* by Sheila Heen, Douglas Stone, et al.

Giving feedback is one of the most important parts of leadership and can and should be one of the most difficult. These conversations can have a huge impact on people's lives, and the organization depends on them. These are high stakes; they can be seen as offensive, even when it is merely a case of differing opinions that evoke strong emotions. I have made it a point to never rush into these conversations. I always meticulously prepare. I always want to ensure that this is a very specific conversation that follows a fixed set of topics and does not digress. Prior to the meeting, I write down the roles, responsibilities, priorities, and expectations to discuss, and I clearly identify the breach, rupture, problem, or incident that did not go well.

After careful, meticulous preparation, I schedule a meeting. In this meeting, I share my understanding of the issue without condemnation or judgment. I then ask the person receiving the feedback to share their understanding of the issue without giving comments until they are finished. I listen carefully and want to hear that they make their own mistakes and learn from them. This is critical to being coachable. If they are willing to own it, then I can ask what they learned and what they are willing to do differently. It now becomes important to review roles, responsibilities, priorities, and expectations, while expressing confidence in their ability to do the job. I then ask if they are willing to put in the effort and work to meet the expectations, but I preface the question by telling them I do not want the answer right now. I express to them that I want them to have time to consider their expectations.

We then schedule a second meeting to talk about their decision, plan actions to correct the situation, and then assess progress. In the follow-up meeting, they need to report back on what they will do to meet their expectations. Then I schedule at least one more follow-up meeting to discuss how they are meeting expectations. Almost everyone I've worked with over the years who was willing to own the situation became a top performer in the organization. The conversation made the relationship stronger and the team stronger because everyone knew and was meeting expectations. In jobs where I was in leadership positions, I've had zero or limited undesired turnover because of these conversations.

Whether you're a manager, a team member, or just an individual looking to make a difference, anyone can be a leader. All it takes is the right mindset, willingness to act, and commitment to caring about those around you. If you want to be a leader, focus on developing your own personal qualities such as empathy, communication skills, and resilience. Take the initiative to tackle problems and find creative solutions. And most importantly, make it a priority to care about the people you work with and those around you.

By doing so, you'll not only be able to make a positive impact on your own life and career, but you'll also be able to inspire and empower others to do the same. That's what true leadership is all about!

Key Takeaways

- Leadership is influence, integrity, value, vision, and inspiration.
- Leadership is not a position, title, or role.

- Leaders set clear expectations.
- Leaders help individuals develop their skills and strengths.
- Trust is essential for an effective team.
- Integrity is the foundation for trust, respect, and leadership itself.
- Leadership is taking ownership of problems, helping find creative and effective solutions, and caring about others.
- Visionary leaders invigorate and motivate an organization.
- Giving feedback is one of the most important parts of leadership.

Triumph

"Every worthwhile accomplishment, big or little, has its stages of drudgery and triumph: a beginning, a struggle, and a victory."

~ Mahatma Gandhi[13]

Joy

I have found joy in the memorable moments of triumph, success, and awe. Joy is biking the Willamette Valley with Scott, climbing a mountain with friends, landing in Brisbane Australia to meet with a customer.

[13] "Mahatma Gandhi > Quotes > Quotable Quotes," Goodreads. https://www.goodreads.com/quotes/37351-every-worthwhile-accomplishment-big-or-little-has-its-stages-of

As you have read, throughout the earlier years of my life, I experienced so few moments of joy and exhilaration. Now I can say that every day is a celebration of some sort; even a simple walk in the woods is special. I cherish relationships, nature, and the triumph of being alive.

Triumph is the outcome of framing your narrative and owning your power. It is the culmination of empowerment, survival, persistence, education, resilience, leadership, self-sufficiency, and purpose, when all the efforts come together and yield success on one's own terms. For me, I feel triumphant when I am helping transform education with technology and when caring for the people in my life whom I love most.

I may never be fully satisfied, as there's always another mountain to climb, professional goals to accomplish, loved ones to support. However, I am more triumphant than I ever imagined, with my perfect (to me) partner, C-level role at the best company in the world (yes, I am biased), and fantastic relationships with my daughters, siblings, friends, and colleagues. This book is my ultimate triumph as I have wanted to tell my story for the benefit of others, and now you are reading the stories and lessons learned. My greatest reward will be hearing from readers about how this book inspires and supports you.

Success

People often asked me to explain what inspired me to work so hard and how I achieved success given all the odds stacked against me. Even when I was in middle and high school, both peers and adults saw the fire burning in me, and I was voted one of the "most likely to succeed." I've told you about many of the people and situations that inspired me, but I didn't want to leave out this special poem that is on my bedroom wall. I read it daily, and sometimes, when needed, multiple times a day. I kept it through college, career, and still have it today. It inspired me and guided my focus. Each line is on a colored strip creating a rainbow effect. The poem, by Barbara Smallwood and Steve Kilborn, is called "Success," and you can find it online. The first line is, "Whatever your mind can conceive and believe, it will achieve." Another line is, "Nothing in the world can take the place of persistence."[14] As I look back at the phrases, I see how my mindset was formed by reading this over and over until it became part of my psyche.

A success-focused mindset enables persistence in pursuit of goals. Progressive goals should be geared toward your purpose. We talked about purpose early in the book, and I want to emphasize here that the achievement of your goals in line with your purpose is success and triumph. Remember, you get to define who you are, how you show up, and your purpose. So, it makes sense that you also create your own definition of success and triumph.

---

[14] "Success," by Barbara Smallwood and Steve Kilborn, motivational hanging wall art 1980S, www.worthpoint.com

Empowered

Winning the 2019 International "Woman of the Year." With me are just a few of the people who helped me achieve this top award, my daughter Briana, me, and my friends Dawne Mayerson and Rocheen Pearson

But this is not as simple as it sounds. We have all been indoctrinated by our families, friends, teachers, and mentors to have a certain idea of what success and triumph mean. It takes self-reflection and conscious experience to unpack and question the overt and subconscious messages we have received that shape our definition of success. Your definition of triumph is personal, and it's yours alone. That is where my three purpose words—contribution, exhilaration, and serenity—became a powerful North Star for me to navigate my life's journey. These are my guiding lights to triumph.

My daughters' graduations were shared triumphs.

I'm going to reiterate this because it is important: *You must define your personal success and triumph.*

Soulmates

When I first was looking for a life partner, I listened to everyone else's advice instead of defining my own version of what I wanted—that didn't turn out well. I realize now that it would have been better to be on my own than with the wrong person. It took me many more years to hear my own voice and know who I am and what is important in a partner. When I quieted everyone else's demands on me, it enabled me to find my true soulmate. What mattered most to me was being supported and seen.

During and after my illness, I was convinced that I was better off on my own, without the life partner that I had dreamed of and prayed for. I didn't want to let my heart be vulnerable again. It was far too painful. I was convinced that there was no one out there for me. I tried and failed to find my true companion so many painful times. My girlfriends were awesome throughout my year-long cancer battle. They came from all over the country to stay with me for a week or more, helping with whatever was needed. I decided I didn't need a man in my life; they just brought heartache and distraction from my purpose.

I became comfortable with plans to travel, go on adventures, and live on my own. I bought an inflatable kayak to transport in my tiny car to any lake of my choice. I joined meetups and booked a ticket with Rhodes Scholars for Spain on Thanksgiving, because it was my year without my daughters.

This old joke comes to mind often: Know how to make G-d laugh? Make a plan. (As a professional project manager, I don't like this joke… especially when the joke's on me!)

My daughter, Briana, had a boyfriend. She wanted to take him to a family event. She wanted to introduce him to our family's unique brand of love, pressure to succeed, plentiful food, joyful celebration, dance, and deep connection. We had a series of conversations where she wondered who would host the next family event.

I told her, "It certainly won't be me having a family event (such as a wedding). There's no one out there for me." She chastised me forcefully, exclaiming, "Mom, don't say that! You can't give up. Get out there, and meet people." Just to prove her wrong, I signed up for an age-appropriate dating site, creating a sassy profile that asked for everything possible from a potential partner—a confident, generous, caring Jewish man who could build and fix things around the house. He would need to have a great sense of humor and be an adventure-seeker like me—someone who loves hiking, biking, skiing, and traveling. To put this mythical creature completely out of reach, I said he had to have a lake house with a boat! I was certain that this "unicorn" did not exist and promptly ignored the app after posting my search criteria.

As fate would have it, I was wrong again. Adam Grant, renowned Wharton professor and author of many wonderful books, taught me to celebrate being wrong. I had previously dated a few interesting men: a movie producer, a wealthy insurance company owner, a politician, several salesmen, and technologists, but none of them were right for me. I pretended to be what I thought they wanted, but it was not my authentic self, and the connection was usually shallow or fleeting.

Then Scott contacted me through this app. After my illness, he was the only one who I met in person, not for a date, just as a potential friend for a concert in the park. He was funny and good-looking. He enjoyed the same activities and seemed more genuine than anyone I had met before. However, he was short. I heard my aunts' and grandmother's voices, "Is he tall?" I questioned that judgment… why did height matter? At first it bothered me, but Scott demonstrated his depth, intelligence, warmth, and incredible ability to see me clearly, love me, and support me. I soon realized that height does not matter. I would love him even if he was short because he is caring, supportive, generous, funny, interesting… simply a *mensch*! I have thanked his amazing mother many times for raising three *mensches* (people of integrity and honor). She should write a book on how to raise strong, sensitive, supportive men.

From what I know, some of the best men are vertically challenged and overcome this by being a better partner!

On our first date, we went to a comedy club and saw Nikki Glaser. Neither one of us had been familiar with her material, but we both loved comedy. She was uncomfortably coarse, as Scott and I were just friends up to this point, and he was a complete gentleman. Her last bit was about dating "out of your league." She commented on how famous males got to date women way out of their league, but for women it just wasn't the same story. A woman could only date out of their league if it was a short guy. She then went on to talk about all the advantages of being a short guy.

I thought Scott must have paid her extra to finish with that bit! We laughed, and this has become a running joke throughout our relationship.

From my experience and self-reflection, I have realized that so many of our built-in judgments about what a partner should look like or do for a living were holding people back from finding a fabulous partner. Too often, as women, we try to be the perfect version of a female that we imagine a man would want. We do our hair and make-up, worry about clothes, diet, exercise, have cosmetic surgery, and purchase 1,000 beauty products to try to become that fake "Barbie Doll" version of arm candy. It does not work; if we find a successful, handsome, tall dating partner but are covering up our true selves, the connection will ultimately fail.

We must be confident in who we are and be living in our purpose before we seek a partnership with someone else and never compromise on what is most important at our core. That is the basis for how Scott and I came together. Cancer stripped my self-image to the bones. I barely had hair, was overweight, and was not "done up" for dating. But Scott saw me for who I was. He recalls that time as meeting the most interesting person he has ever known and still thinks so today. He accepted me for who I am. He loves and supports the good, bad, ugly, my drive, my long fuse but intense temper, long work hours, and my unconditional love for my daughters.

Over the next few months, we got to know each other better, and our bond deepened. Scott clearly had a wonderful personality and character, but he let me in on his flaws and attributes over time. He is incredibly smart, with multiple B.S.'s and multiple master's degrees. He is super handy — having rebuilt his lake house beautifully, in the same aesthetic as my house, coincidently, after stripping it down to the studs. He has so many talents and interests.

He's been a certified open water scuba diver since he was 15, a licensed commercial pilot, licensed skydiver, a former restaurant owner, used car dealership owner, and a successful IT consultant. His abilities seem endless, and he never stops trying new things, growing, and learning. He had worked in the family taxi and limousine business since he was six years old, beating me by six years! We both have the same intense work ethic, but he has endless energy, like I used to have. In his spare time, he does the laundry, dishes, cooking, hundreds of pushups and core exercises, and home repairs. The pandemic, in addition to all the stress and challenges, was a gift for us to be together. We never tire of each other's company, hugs, and love.

Scott is incredibly sensitive, which has a double edge. I love how he is emotionally open and communicative, but it means I must be careful not to say and do unintentionally hurtful things. We took a couples communication course (Imago) early in our relationship, and we make use of the tools we learned regularly. Scott just has to say, "Can we dialogue?" and I know he needs to express something important.

Scott came into my life, and we found love without drama or trauma. He is so easy to be with and supportive. We laughed and played like I never had before. We proved that it's never too late to have a good childhood! The best life triumph for me has been meeting and marrying my soulmate. We recognized our deep and layered connection within a few weeks of our first date and married less than a year later. It's only grown more incredible. We both love adrenaline-producing activities: lakes, mountains, travel, and so much more. He is the love of my life, and every day is a celebration as we experience our love, joy, and companionship. He is the perfect partner for me. I imagined and manifested my true companion.

Serena Sacks-Mandel

We talk about our "Love Language," and for me, it's "acts of service" and "quality time," which Scott loves to do. He says that love is a series of intentional actions. Every morning, he feeds my cat and then brings me a hot cup of tea just the way I like it.

Through this act of service, I know that he loves me, on that day and every day.

The basis of our relationship is striking and noteworthy. We are first and foremost best friends, we are partners in all decisions and have each other's back, and we are present to meet each other's needs. We laugh together and share our lives together. There are no walls, no pretenses, and no barriers. We both seek the next adventure, whether that is career, travel, or building a dream home. There are so many belly laughs, and at times, we laugh until it hurts.

We are in-tune with each other and feel it when we get out of tune. It makes us giggle the number of times a day when we are thinking the same thing at the same time. Even doing simple things, we tend to act more like one person with four hands and two heads working in concert. We still surprise each other when we reveal a hidden truth and find the other had the same truth. We marvel at how similar we are, even though we grew up under such different circumstances and half a country apart.

I shared all my truth with Scott before we married. I wanted to give him the choice to run away quickly before things got intense. He was not afraid to see me with all my faults and my true self. He is the other half of me that gives me balance. He makes me feel whole, complete, loved, validated, and supported in every way. Now I know what true partnership is and understand why previous relationships were never going to work. It is very different when you are with the right person: someone who cares deeply about your happiness as I do his, who enjoys doing many of the same things together, and knows you better than you know yourself at times. Most importantly, the right person is fun to be with, and you enjoy laughing together often.

We met just prior to the pandemic and braved it together, and now we live in the post-pandemic world, which has allowed us to avoid long commutes and focus our work time and personal time. Since March 2020, when the pandemic started in the United States, I have been working 60 to 80 hours a week with 10 to 12 hours a day in live video meetings. I must set boundaries and claim my time, but I feel now I have much more control over it. I am more careful about not checking emails or logging in on Saturdays and ensuring I get enough sleep and exercise.

Through my marriage to Scott, I gained another daughter, Chelsea. She is easy to be with, and I am proud that she is studying computer science and living with us. I treasure the time we spend together. It's a joy to see her grow in self-confidence and enjoyment of life. She treated me to breakfast for Mothers' Day this year.

Empowered

From Resilience and Strength to Triumph

One of my dearest friends, mentors, colleagues, and former boss has an awesome story of resilience, strength, and triumph. She has inspired me to be a better leader, daughter, and mother. We took care of each other's children when the other had customer meetings, and she stayed with me for a week when I was sick, watching me suffer in pain while the nurse tried multiple times to put the needle in my chest port without any numbing agent.

I met Diane while she was working for a firm contracting to Disney, and I was a manager at Disney. Arielle had just been born, and she was pregnant with her third son. She remembers me as intimidating; I remember her as smart, savvy, and personable. Not too long afterward, I went to Harcourt, and she called on me again, this time after the baby was born. We immediately recognized each other as kindred souls. We were both tough, smart, and driven women in tech. I needed her help at Harcourt, and she understood Harcourt's resource needs at a much deeper level than just the technical skills on a resume. She helped us find smart, forward-thinking candidates who had positive energy and were adept at adapting to new environments.

Diane was invited to my 10-year anniversary party and arrived without her husband. I asked why, and she promised to tell me later. It turned out she was getting divorced; one of the first in my friend group. I was sad for her and then incredibly impressed. She was strong enough to ask her husband to leave, even with three young children. She took care of them and raised them on her own. Their father never contributed financially, yet she stayed friendly with him, allowing him to stay at her home for visits. While raising these boys, she started her own consulting firm, which I worked for at times, helping her recruit talented consultants and leveraging their expertise in several roles.

She was an inspiration to me after I divorced as she had found true love with Mike, a wonderful man and loving husband. She has also been an incredibly caring and attentive daughter to her aging parents who have now passed. Diane is an all-around good person. Not only is she a good person but also a wonderful businesswoman who has been recognized with many awards. She has been a leader at the Juvenile Diabetes Research Foundation (JDRF) and other nonprofit organizations, as well as a supporter of high school girls with limited means and high potential through her program Me and My CommUNITY. She is a loving wife, mother, friend, and daughter. Diane embodies all the characteristics in this book, and her story is a true triumph.

For me, triumph is less about awards and recognition. It's about how I've helped others reach their moment. Student graduations as well as my own daughters' are great examples.

Empowered

Both of my daughters had many challenges on their road to graduation, and it was a joy to see them triumph.

Serenity

Serenity, that aspirational state, happens more often now than it used to. The first picture was the first hike Scott and I took together in Tennessee. He brought me to his favorite waterfall, which is like a rock bowl. You hike to the bottom and then look up. I was doing a yoga pose and a rainbow appeared over me. It was *bashert*, meant to be. This was probably the moment I fell in love with him. The next one is on a truly relaxing and romantic vacation. The last is on our boat on the Tennessee River, one of our many sunset cruises that are truly peaceful and serene. It is a triumph that serenity is attainable now.

I have learned from and enjoyed every phase of my journey, no matter how challenging. The best is still ahead of me though. I feel triumphant. I have overcome so many obstacles, healed so many wounds, and persevered through challenges. I've practiced resilience and aligned my actions with my purpose. As a result, I have been recognized many times over as Woman of the Year, CIO of the Year, Technologists who Matter, Education Visionary, and others.

I have the benefit now of being a leader and enjoy coaching, mentoring, and advising other individuals and organizations. I am still working at Microsoft full-time; I continue to sit on boards and support nonprofit organizations that matter to me. I can see my next great adventures are still ahead of me. Hopefully, several years from now, I look to be in my next chapter when more of my time will be spent speaking, consulting, coaching, advising companies, board work, and of course, family and leisure.

Healing Relationships

Love is also about doing the tough work of communicating, apologizing, and making things better. If you have a daughter, you may be familiar with teen schism. They separate from us and can be mean, hurtful. You may be frustrated and angry with them. There are endless coming-of-age TV shows and movies based on the teen drama and trauma of these years. These were difficult years for us, but we went through them together and now have a stronger bond.

Empowered

My experience with each of my daughters, mostly due to my poor handling of the divorce, was particularly intense and painful for all. I deeply questioned our ability to recover and heal from the bitter arguments. Yet, my daughters and I are extremely close and supportive of each other now. They are my core people, and I am just thrilled to be a part of their lives. I am in awe of their wonderful accomplishments, their ability to care for and about others, and their incredible prospects. We have worked incredibly hard to repair the damage. I've done a lot of apologizing and am still doing EVERYTHING possible to demonstrate my love, after asking them to tell me what they want and need from me as their mother. I had to quiet my inner voice and just listen to how I hurt them and what would make it better. It has taken many years of remorse, honest communication, and giving of myself. I still have work to do for them to see I can be who they want me to be for them. I am now seeing them trust me again and feel that they love me, a triumph.

Listening intently and hearing their needs, putting their needs above my own needs, and being patient and consistent with my love and support has worked for me in re-building a relationship with them. I must be warm, loving and open to owning my part and moving forward together. A note I received from my daughters meant so much to me. They see me and love me.

I have seen that many daughters blame their mothers for things they don't like about themselves as they define themselves and become young adults—so a natural separation occurs. Of course, there are exceptions to this situation, but I have seen many mothers and daughters fight over the years and eventually come back together. When I was in the throes of this phase, my friends assured me that it would work out as they grew up. At the time, I couldn't see it, but they were right. I gained a deeper understanding of my mother when I had children of my own. It will complete the circle and be a triumph if one day my daughters experience the same.

Empowered

One of the greatest triumphs in my life is the healing of my relationships with my daughters. I created harm, chaos, and confusion when I only wanted to love and protect them. As I was writing this book, we had a family Thanksgiving with my daughters, their cousins, and my siblings. It was wonderful to have everyone together. We visited with my mother, and that family reunion was amazing. After all my family has been through, coming together for a holiday is a triumph.

To bring my story full circle, I now know that I am supposed to be here. I have helped each one of my siblings, husband, children, and friends to realize their purpose and potential. My relationships with my daughters, siblings, husband, and all my friends are my triumphs.

Me, Norman, Cheryl, and Larry (my siblings)

Key Takeaways

- Triumph is the outcome of framing your narrative and owning your power.
- Triumph is the culmination of empowerment, survival, persistence, education, resilience, leadership, self-sufficiency, and purpose.
- Pursue your own definition of success.

Postscript

I want to give credit to Brené Brown who is my idol. I am her biggest fan. It was such a thrill meeting her live online. She wished me a happy birthday two years ago on a team call. I love that she said, "There is no joy without gratitude," and I have found this to be so true.[15] The more we are grateful for what we have, the more we find joy in life. Her next quote is much of what I try to convey with this book: "You either walk inside your story and own it, or you stand outside your story and hustle for your worthiness."[16] Take it from me — the latter is never a good place to be.

Own your power. Frame your narrative. Don't sell it, give it away, or disregard it.

---

[15] "Brené on Strong Backs, Soft Fronts, and Wild Hearts," BreneBrown.com, Nov. 4, 2020, https://brenebrown.com/podcast/brene-on-strong-backs-soft-fronts-and-wild-hearts/

[16] "In You Must Go: Harnessing the Force by Owning our Stories," BreneBrown.com, May 4, 2018, https://brenebrown.com/articles/2018/05/04/in-you-must-go-harnessing-the-force-by-owning-our-stories/

My family journey and my work journey are still a work in progress. It has taken until my current role at Microsoft for me to see myself as an employee of an organization instead of seeing my job as my identity. My story has been a culmination of my cancer story, my love story, and maturity. I am a product of history and how I framed my narrative. It is a victory for me to be alive, a wife, mother, daughter, friend, and whole person rather than the embodiment of a job in an organization. Before I truly found my purpose and my love and grew older, my entire being was focused on work.

If I can help just one person learn from my journey and accelerate their path to wholeness, become more conscious of all the elements of their life, and find their own value in each, then this process of writing and sharing this book will be a triumph. The main benefit for my learning and growth is that it contributes to others.

Key Takeaways from *Empowered*

- Choose a partner who enjoys the lifestyle that makes you happy.
- Lifelong learning is a must.
- Find your purpose and persevere.
- Live your values.
- Assume goodwill; most people are not trying to hurt you.
- When needed, ask for help from your friends, they will be grateful to be of service.
- Believe in abundance instead of scarcity and hope instead of fear.
- Don't wait for your life to be at risk to discover who you really are.
- Capture joy along the way.
- Seek to learn and add value, every day, in every role.
- Treasure your friends and let them know how much you appreciate them.
- Focus on gratitude and forgiveness.
- Own your power. Don't give it away in jealousy, anger, hurt, or sadness.
- Process pain and anger (natural emotions), and then move through them and let them go.
- Always make changes that move you closer to your purpose, mission, and goals.
- Let go of those who hold you back or tear you down.

Empowered

- During challenges, keep in mind that crucible moments shape us and guide us to move forward.
- As you overcome obstacles effectively, you will be invited to take on larger responsibilities.
- Every day is a job interview; everything you do advertises your personal brand.
- Lead with humility and courage, regardless of your title.
- Remember, it's your life, and you deserve to be happy (hint: happiness comes from contributing to others).

Recommended Books:

- *Braving the Wilderness*: Brené Brown
- *Dare to Lead*: Brené Brown
- *Atlas of the Heart*: Brené Brown
- *The Gifts of Imperfection*: Brené Brown
- *Give and Take*: Adam Grant
- *Think Again*: Adam Grant
- *Hidden Talents*: Adam Grant
- *What Got You Here won't Get You There*: Marshall Goldsmith
- *How Women Rise*: Marshall Goldsmith and Sally Helgesen
- *Culture Code*: Daniel Coyle
- *7 Principles for Making a Marriage Work*: John Gottman
- *Too Good to Leave, too Bad to Stay*: Mira Kirshenbaum
- *5 Dysfunctions of Team*: Patrick Lencioni
- *The Motive*: Patrick Lencioni
- *The Ideal Team Player*: Patrick Lencioni
- *Emotional Intelligence* (Series): Daniel Goleman
- *The Infinite Game*: Simon Sinek
- *Start with Why*: Simon Sinek
- *Blended*: Michael Horn
- *From Reopen to Reinvent*: Michael Horn

- *Thanks for the Feedback*: Douglas Stone, Sheila Heen
- *The Funny Thing About Forgiveness,* Andrea Flack-Wetherald
- *Getting the Love you Want*: Harville Hendrix and Helen LaKelly
- *Keeping the Love you Find*: Harville Hendrix

About the Author

Serena Sacks-Mandel is the worldwide chief technology officer at Microsoft for the education industry. Prior to joining Microsoft, she was the chief information officer at two unique technically advanced large public school districts where she enabled student-centric teaching and learning, which resulted in significant improvements in student outcomes.

Prior to pivoting to education, she led innovation and transformation teams at several large companies and provided management consulting at many others. Serena has won numerous state, national, and global awards for leadership, vision, technical excellence, and her commitment to supporting women in technology.

In her non-working time, Serena and her husband Scott are adventure seekers—biking, hiking, traveling, and boating, often with their three wonderful adult daughters.

Printed in the USA
CPSIA information can be obtained
at www.ICGtesting.com
JSHW012123120324
59039JS00004B/18

9 781958 481233